THE
BRIGHTEST STARS

DISCOVERING THE UNIVERSE
THROUGH THE SKY'S MOST BRILLIANT STARS

Fred Schaaf

WILEY

John Wiley & Sons, Inc.

This book is dedicated to my wife, Mamie,
who has been the Sirius of my life.

Copyright © 2008 by Fred Schaaf. All rights reserved

Published by John Wiley & Sons, Inc., Hoboken, New Jersey
Published simultaneously in Canada

Illustration credits appear on page 272.

Design and composition by Navta Associates, Inc.

For general information about our other products and services, please contact our Customer Care Department within the United States at (800) 762-2974, outside the United States at (317) 572-3993 or fax (317) 572-4002.

Wiley also publishes its books in a variety of electronic formats. Some content that appears in print may not be available in electronic books. For more information about Wiley products, visit our web site at www.wiley.com.

Library of Congress Cataloging-in-Publication Data:
Schaaf, Fred.
   The brightest stars : discovering the universe through the sky's most brilliant stars / Fred Schaaf.
       p. cm.
   Includes bibliographical references and index.
   ISBN 978-0-471-70410-2 (cloth : alk. paper)
   1. Stars—Luminosity function—Amateurs' manuals. 2. Stars—Amateurs' manuals. 3. Astronomy—Amateurs' manuals. I. Title.

   QB815.S33 2008
   523.8–dc22

                                                                        2008000278

Printed in the United States of America

10 9 8 7 6 5 4 3 2 1

CONTENTS

ACKNOWLEDGMENTS

The first person I want to thank in connection with this book is Kate Bradford. Kate acted as acquisitions editor for two books of mine at the same time: *The Brightest Stars* and *The 50 Best Sights in Astronomy*. Now that both books have come to fruition (*50 Best Sights* was published by John Wiley & Sons in 2007), I feel extremely gratified to have been able to bring them into being. But the whole process could not have gotten off the ground without Kate's skilled help and support.

The next person I worked with on these two books was editor Teryn Johnson. I've not forgotten her congenial support. The person who has worked the most, and the most vitally, with me on these books, however, has been Christel Winkler. She has been patient and understanding under trying circumstances. I wish to give my deepest thanks to her for her tremendous and conscientious efforts to keep these books on schedule.

Now let me turn to the diagrams, maps, and artwork produced for *The Brightest Stars*. Many of them were created by two old friends of mine, Guy Ottewell and Doug Myers. Their work is always unique and brilliant. I owe an enormous debt of gratitude to both of them.

Vital maps were also provided by Robert C. Victor and D. David Batch, who produce the Abrams Planetarium Sky Calendar. Sky Calendar is a wonderful resource for all knowledge levels of skywatchers, and teachers, too. It is available from the Abrams Planetarium, Michigan State University, East Lansing, Michigan 48824. You can also check out the associated Skywatcher's Diary at www.pa.msu.edu/abrams/diary.html.

INTRODUCTION

The season was winter. It was probably the winter I turned six years old. All I know for sure is that things got started when I read a section of a book at school. The section told about the brightest stars and constellations of winter. That night, I lay awake in my dark bedroom, peering out through the slits between the big window's venetian blinds. And that was how I first really made contact with one of the greatest inspirations of my life: the stars.

But it was not just any stars I was seeing through the blinds and outside between the bare winter tree branches. These were the brightest stars. First, the blue-white star Rigel, which the book had talked about, glittered through the trees and truly pierced my heart with its beauty and wonder. Only I didn't know initially that it *was* Rigel. It was so splendid I thought it must be Sirius, the blue-white brightest-of-all star that the book had also told me about. It wasn't until minutes later, as I lay watching and pondering Rigel, that I suddenly caught sight of a second spark of blue radiance. My already wonderstruck mind reeled because this blue light burned several whole qualitative levels of brightness greater than the first star. I was gazing, for the first time, upon Sirius, star of stars, the brightest of all stellar jewels in the firmament. And I would never be the same again, for I had started to make the acquaintance of the brightest stars.

Welcome to a book devoted to the brightest stars. Much has been written about the planets of our solar system, and rightfully so, for they are marvelous. But when we look up in the sky we see twenty-one stars that are usually brighter than Mars. One of them—my childhood flame, Sirius—is almost always brighter than any of the planets except Venus and Jupiter. And these stars do not shine with the steady light of the planets. Stars twinkle. That twinkling, although it is caused by Earth's atmosphere, beckons us to beyond— beyond our planets, beyond this solar system, beyond the reach of our spaceships for the foreseeable future, the stars call to us. And the message of starlight is first that these pretty pieces of trembling fire are suns in some essential ways like our own. Even without close inspection, we can realize that we are seeing suns and imagine that some, perhaps many, of the stars have their own sets of planets. We can even imagine that maybe sentient beings like ourselves are staring back and wondering about the stars in *their* sky, including our own Sun.

The brightest stars deserve a book of their own. Stars are the most important units of the universe at large. The very word *astronomy* means "ordering of the stars." When we look up in the natural night sky, almost every one of the multitude of lights we see is a star. It occurred to me in a flash a number of years ago that a great way to teach people about stars in general was to teach them about the most outstanding individual stars, using those stars as powerful exemplars. One could do this by selecting the most technically interesting stars, even ones so distant or hidden that the largest amateur telescope could never show them. But the best form of learning is learning through direct involvement—in astronomy, through actual observation. This is especially true in today's world, in which a vast majority of people live under skies significantly degraded by human-made light pollution. In such a world, it is the *brightest* stars that everybody can see, even people who have only their unaided eyes to use.

Are all the major kinds of stars represented by the twenty-one brightest, the stars of the so-called 1st magnitude? Not quite. But if you count the red dwarf and white dwarf companions of these stars, most of which are within the reach of amateur telescopes, you really do have a nearly complete representation of the basic different spectral types, luminosity classes, and special categories (double stars, variable stars) of stellar bodies.

The brightest stars have not had a book really devoted entirely to them for many decades, perhaps not for a century, since Martha Evans Martin's delightful 1907 work, *The Friendly Stars*. Martin's book was aimed mostly at the completely novice stargazer. The book you now hold in your hand is really meant for a wider range—everyone from the absolute beginner to the veteran amateur astronomer.

The opening chapters of this book tell and show how to find the brightest stars, along with some of the major constellations of the sky. Those chapters also discuss many of the basics of astronomical observing. They are followed by a section whose chapters explain the fundamentals about the nature of stars and their various kinds. An experienced amateur astronomer may find these chapters a good refresher course in stellar astronomy, hopefully gaining some new insights by original ways I've tried to present the information.

But it's the third section of this book, which forms more than half the book's length, that will, I hope, be largely original to the vast majority of readers. Part three is devoted to chapter-by-chapter profiles of each of the brightest stars in the sky. Each profile gives detailed information about what to look for in the observational experience of the star, including the star's environment of season and constellations, of earth and the sky. Each profile also offers the lore and legends connected with the star, for these are a measure of what the human race as a whole has found interesting and individual about the star.

A key part of each profile, of course, is also what the science of astronomy has taught us about the physical nature of these brightest stars—that is, what they are like as suns. Much of this information is shockingly new. About ten years ago, the positional and brightness measurements of the Hipparcos satellite helped begin a revolution in our knowledge by allowing us to determine with far greater precision the distances to many stars. Once this was known, we could determine the true brightness of them. Combining this with what the spectra of stars tells us about their chemical composition enables us to truly understand them as individuals. Other recent technological breakthroughs have also helped greatly.

In the past five years, what astronomers have learned about the familiar brightest stars in our sky is amazing. They've revealed evidence of planets around three of the twenty-one stars (with more revelations of this sort no doubt on the way). They've made major discoveries about the age and fate of many of these stars. They've solved a long-standing mystery of why one of these stars is much brighter and hotter than theory would predict. They've found out that another star is only two-thirds as far from us as was previously believed. In these past five years, astronomers have finally determined the strange shapes of some of the brightest stars and how fast they rotate (some spin dozens of times slower than our Sun; others, a hundred times faster). Astronomers have also learned some of the spin orientations, with amazing results: one of the brightest stars flies past us perfectly sideways like a spinning bullet, whereas another points a pole right at us, so that our Sun must be the polestar in its sky. Only in recent years have scientists also been able to determine with accuracy where the brightest stars have been in the past few million years and will be in the next few million. This enables us to know which stars have been the very brightest in our sky into the distant past and the distant future, and to uncover some individual spectacular surprises. For instance, we've learned that two of our brightest stars were even brighter in the past at a time when they passed close to each other in both our sky and in space.

This book obviously attempts to do many things. But what would I say is its ultimate goal? That is very simple. The goal is to get you, the reader, out to meet the brightest stars for yourself. I hope you'll start looking at them even as you start the journey through the book and return to the chapters on the individual stars many times when you observe those wondrous luminaries once again. The stars have always been a symbol of humanity's highest aspirations. And the brightest stars are the chief expressions of the wonder and magnificence of the stars. Surely there could be few things in life as uplifting as meeting them. I can do no better than follow the classic star-name expert R. H. Allen in quoting the words of early American poet Lydia Sigourney: "Make friendship with the stars!"

PART ONE

STARS
IN THE SKY

How Bright Is Bright?

This is a book whose central figures are the brightest stars. All of us under-stand the basic sensation and idea of "brightness." And it is to be hoped that even in this age of widespread urban light pollution and separation from nature, all of us have seen a star at one time or another.

But which are the brightest stars? Before we know which stars to focus on as the brightest, we need to define what we mean when considering a star to be one of the brightest.

A Sky Full of Stars—and Several Questions

The first step we take is out from beneath the trees on a clear, moon-free evening. We are many miles from city lights and, over the field where we stand, there is nothing but stars: stars by the hundreds and seemingly thou-sands, stars bunched here and lined up there, stars that seem alive with twinkling and are vivid with radiance in the depths of the velvet dark sky.

But we immediately notice something else about the stars: how mar-velously many different degrees of brightness they come in. The dimmest stars are too numerous to count, in every direction. A few hundred modestly bright stars are sprinkled generously among the multitude of fainter ones. Of the few dozen even more radiant stars, each seems to dominate—sometimes with a similar partner—its own small section of the sky. And then there are the brightest stars. True flames of prominence in a dark country sky like this, as few as five of them may be in the entire sky at a single time. But each rules its entire direction and large realm of the starry heavens.

These brightest stars are the outstanding examples of stellar prominence, interest, and beauty. They are brilliant enough to detect even in the midst of glaring city light pollution. But a look at a clear night sky in the country (or even, to a lesser extent, from small cities and suburbs) shows us enough lev-els of stellar brightness that we have to ask several questions: Where do we draw the line to end the list of "brightest stars"? How much brighter than the faintest stars visible to the naked eye are the brightest? What we need to

The magnificence of a bright star. This one is Regulus. The image also shows the Leo I dwarf galaxy.

know, most basically, is how astronomers measure and rate brightness. In other words, how bright is bright?

A Matter of Magnitude

The concept and term that astronomers use most often to measure brightness is **magnitude**. The term was first applied to the brightness of stars by the ancient Greek astronomer Hipparchus (c. 120 BC). Hipparchus recognized six classes of star brightness. The brightest stars were said to be "first magnitude"—the first, brightest class. Slightly less brilliant were stars of "second magnitude," then "third magnitude," and so on with decreasing brightness. The faintest class of stars, those just barely visible to the naked eye on a clear night, were said to be "sixth magnitude."

How many stars belonged to the first magnitude of Hipparchus? Skywatchers decided that the stars ranging from the single brightest, called Sirius, down to the fifteenth brightest, called Regulus, were the ones that should belong to this category. This included Pollux, the brightest star of the constellation Gemini, but just barely excluded the star Castor, the second brightest star of Gemini, located very close to Pollux. Pollux was considered a 1st-magnitude star but Castor only a 2nd-magnitude star.

The Hipparchus system of magnitude worked well until modern times. Then astronomers needed to refine and more precisely quantify the magnitudes of stars and other celestial objects. In 1856, Norman R. Pogson noted that the radiance of a typical one of the brightest stars is about 100 times stronger than that of a typical one of the faintest stars visible to the naked eye. He therefore suggested that a difference of five magnitudes (for instance, the difference between a star of exactly magnitude 1 and a star of exactly magnitude 6) should be set as equaling precisely this ratio of 100 to 1. And speaking of precision: We can use fractions or, better yet, decimals to specify exactly how bright a star is. A star of magnitude 1.0 is exactly 100 times brighter than one of magnitude 6.0. But we could also speak of a star whose magnitude is 1.2 or 5.4 (or 2.6 or 4.8, and so on). A star of magnitude 1.5 would be halfway in brightness between a star of magnitude 1.0 and 2.0. The star of magnitude 1.5 would be brighter than the one of 2.0. Always remember: the *lower* its magnitude, the brighter an object is.

Pogson's system of magnitude has been adopted by astronomers around the world. That difference of 100 times equaling 5 magnitudes seems convenient indeed. The only problem is that it means a difference of 1 magnitude has to be the fifth root of 100—that is, the number that, when multiplied by itself 5 times, equals 100. But the fifth root of 100 is 2.5118864 . . . . So a star of magnitude 1.0 is 2.5118864 . . . times brighter than a star of magnitude 2.0. Isn't this number awkward? For use by the average sightseer of the heavens, yes. But, of course, professional astronomers are mathematically adept enough to have no problem manipulating such a figure.

Extending the Old Range of Magnitudes

Modern astronomy not only needed precise and agreed-upon values for magnitudes; it also needed to extend the old range to include objects dimmer than sixth magnitude and objects brighter than first magnitude.

First, consider higher (dimmer) magnitudes. Telescopes can reveal stars and other celestial objects much dimmer than the magnitude 6.5 often suggested as the faintest visible to the average unaided eye in clear, dark country skies. The limit for a pair of binoculars with 50 mm (2-inch) lenses might be about magnitude 10; for a telescope with a 6-inch-wide mirror or lens, about magnitude 14; and for some of the world's largest telescopes, perhaps magnitude 20—about a quarter-million times dimmer than can be glimpsed by the typical naked eye. The Hubble Space Telescope, using electronic imaging and its longest exposures, has recorded galaxies as faint as magnitude 30—2,500,000,000 (2.5 billion) times dimmer than the naked eye.

Now let's turn to the opposite end of the magnitude scale. Many celestial objects are brighter than a standard 1.0 star. First, astronomers have found that no less than fifteen stars are brighter than 1.0. These have magnitudes like 0.87 and 0.45. (With modern photoelectric measuring devices, it is possible to come up with magnitudes for the brightest stars that are probably no more than 0.03 magnitude in error, at most.) But four of the fifteen stars brighter than magnitude 1.0 are even brighter than magnitude 0.0. For their brightness, we have to employ negative magnitudes like −0.3 and −0.62. In fact, the very brightest of night's stars burns so brilliantly it exceeds −1, glowing at −1.44. Venus, the most brilliant planet, can shine as bright as −4.7. The Full Moon floods the night with a radiance of −12.7— about 40,000,000 times brighter than the faintest naked-eye star. And the Sun, blindingly bright, creates day when it is above the horizon, with a magnitude of −26.7—about 1,600,000,000,000 (1.6 million million or 1.6 trillion) times brighter than the faintest star we can see with the unaided eye. (Remember that a difference of five magnitudes is 100, and ten magnitudes is $100 \times 100 = 10,000$.)

The Full Moon is more than 11 magnitudes brighter than the brightest star, or about 30,000 times brighter. But all the light of the star is concentrated into a point of light—so its light is much more *intense* than that of the big Moon.

Defining 1st Magnitude

The old dividing line between a 1st-magnitude star and a 2nd-magnitude star, you recall, was felt to fall at a brightness fainter than Pollux and Regulus but brighter than Castor. Modern measurements of these stars' brightness supports this, if we consider all stars between magnitude +0.5 and +1.5 as being 1st magnitude and all between +1.5 and +2.5 being 2nd magnitude. Pollux burns at magnitude 1.16, Regulus glows at magnitude 1.36, and Castor shines at magnitude 1.58. All three of these stars are very famous, partly from being stars in the zodiac (the renowned band of constellations we'll discuss in chapter 3). But there is actually one far less famous star that is intermediate in brightness between Regulus and Castor. That star is Adhara, whose magnitude is 1.50. Does that make Adhara the faintest 1st-magnitude star? No, first magnitude is taken as extending to but not including 1.50. Adhara is always considered a 2nd-magnitude star (albeit the very brightest 2nd-magnitude star possible). Adhara has not gained the widespread attention of Castor because it lies rather far south and in Canis Major—a constellation that is absolutely dominated by the brilliance of the brightest of all stars, Sirius.

The old tradition was to consider Regulus and all stars brighter than it to be 1st-magnitude stars. But, if we want to be technical and insist that only stars between 0.5 and 1.5 are of 1st magnitude, we have to assign other magnitudes to stars that are even brighter. The eight stars that shine between −0.5 and +0.5 can be called "zero-magnitude stars." The two stars between −1.5 and −0.5 can be called "−1-magnitude stars."

In common practice, however, astronomers don't usually speak of zero-magnitude and −1-magnitude stars. When someone today speaks of the 1st-magnitude stars, he or she usually means all the stars brighter than 1.5—now, as in ancient times, the stars from Sirius to Regulus in rank. This is the convention we will follow in this book.

The Brightest Twenty-One

There is, however, an addition to the ancient inventory of 1st-magnitude stars, which we must incorporate. The ancient number of 1st-magnitude stars from Sirius through Regulus was, as I said above, fifteen. But these were only the stars brighter than magnitude 1.5 that were visible from Greece and Rome. There are a total of six additional stars that fit that brightness criterion but are visible only farther south in the world.

Altogether, then, the ranks of the 1st-magnitude stars is now known to include twenty-one stars. These are the ones, great in their brightness, lore, and scientific interest, that are the central figures of this book.

MEET THE 1ST-MAGNITUDE STARS

2

In chapter 1, we discussed how astronomers measure brightness in terms of magnitude. We also established that the brightest stars—the ones we will focus on in this book—are those of 1st magnitude.

Now it is time to for us to meet the individual 1st-magnitude stars. Our initial encounter with them here comes as a swift survey of the bright stars of each season.

Of course, if you want to truly observe all these stars, to do so will require a year, or the better part of one. It would be time well—in fact, gloriously well—worth taking. Indeed, throughout this book I shall be emphasizing the importance of observation and direct experience. It is vital (in both the figurative and the literal senses of the word—"key" and "living") to go outside and actually see the stars. Or, better yet, to see and experience them.

The latter is what you will do when you meet each star in the greater context of weather and the environment, even holidays and the clothing you wear, which you find yourself in during a particular season.

For our present tour, let's assume that we have only untrained naked eyes. As long as we also assume that we have for each season a night of clear air and Moon-free country sky, we will be able to enjoy the full splendor of these stars.

The 1st-Magnitude Stars of Winter

Winter is the greatest season for bright stars. They blaze so prominently on cold nights that even many people who almost never glance at the sky will tell you something they've noticed: "The stars shine brighter in winter." The conclusion that most people draw is that skies are clearer in winter and that this is why the stars have that special extra luster.

In reality, most lands have both their least cloudy sky and most transparent (unhazy) air at other times of year. Early autumn is best in these respects in much of the United States and Canada.

The true reason "the stars look brighter in winter" is that they *are* brighter: on winter evenings, there happen to be more bright stars in the

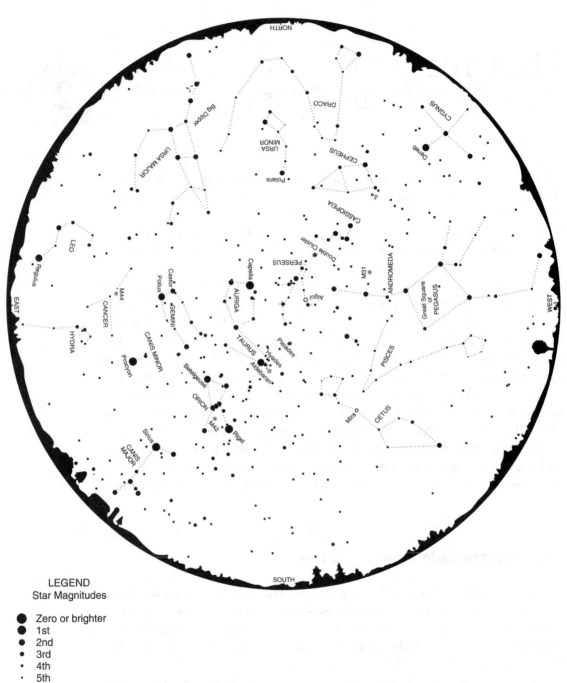

LEGEND
Star Magnitudes

● Zero or brighter
● 1st
● 2nd
● 3rd
· 4th
· 5th
✳ Deep Sky Objects

January mid-evening sky (December, late evening; February, early evening).

directions of space that are then presented to the Northern Hemisphere of Earth. Or we should say in one direction: either east, southeast, or south (depending on how late in the night and how far along in the winter you look). For across a relatively small span of heavens in winter, there shine no less than seven of the fifteen 1st-magnitude stars easily visible north of the tropics. And they sparkle in at least six of the brightest of all constellations.

The single brightest constellation of any season is Orion, the Hunter. On a winter evening you should almost immediately notice him and his striking Belt of three 2nd-magnitude stars in a short, straight row. The most conspicuous star cluster of any season, the lovely, tiny, dipper-shaped Pleiades, is also a sight that instantly draws your attention at this time of year. Of all naked-eye sights beyond our solar system, there is only one that catches our gaze as quickly as Orion and the Pleiades—more quickly in twilight or moonlight, where its greater brilliance dominates. Like them, it shines on winter evenings. It is the brightest of all stars, Sirius.

Sirius! It is not just the brightest nighttime star visible from anywhere on Earth (or anywhere in our solar system, for that matter). It is *by far* the brightest star, appearing at least two qualitative levels more brilliant than any of the other radiant winter stars. Orion's Belt, which typically lies well to the upper right of Sirius, roughly points to it. But when it is visible in the sky, there is no mistaking Sirius. A few of the planets outshine Sirius but they do not twinkle—and in Sirius, this twinkling is often enhanced beyond mere prettiness to a magnificence that can be literally breathtaking. This is especially true when Sirius is low in the sky, as well as when it is higher but our atmosphere is unusually turbulent. In either case, you will see Sirius not just throb and dance with twinkling but also pulse with bursts of all colors. In no other star are twinkling and color changes anywhere near so prominent to the naked eye.

Sirius is several times brighter than any of the other six 1st-magnitude stars of winter, but the others are captivating both in their own right and within their respective constellations.

Orion's two 1st-magnitude stars are blue-white Rigel and golden orange Betelgeuse (almost always dimmer than Rigel but variable in brightness, so that once in a great while it does outshine Rigel). Rigel and Betelgeuse lie diagonally from each other in the rectangle or hourglass that forms Orion's body, with the three-star Belt between them.

Upper right from Orion (pointed at by his Belt in the opposite direction from Sirius) glows slightly orange-gold Aldebaran. Aldebaran gains much prominence from being the end of one arm of a V shape of stars. This V (or arrowhead shape) represents the face of Taurus, the Bull, and, except for Aldebaran, is in reality the Hyades star cluster. Aldebaran marks the Bull's

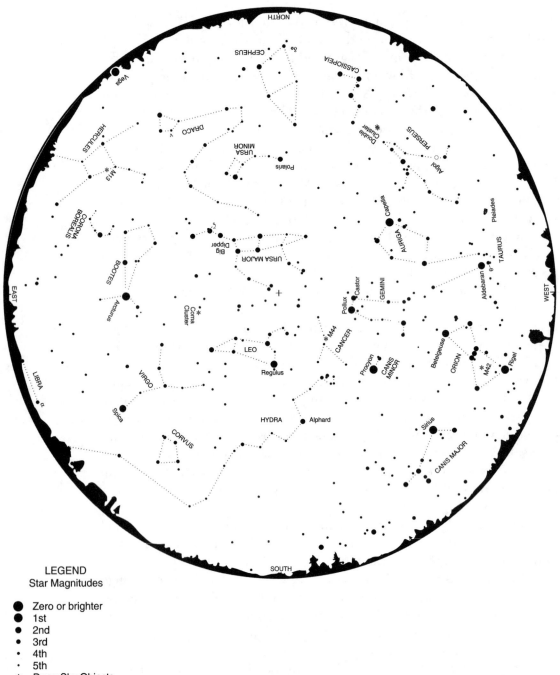

LEGEND
Star Magnitudes

- ● Zero or brighter
- ● 1st
- ● 2nd
- ● 3rd
- · 4th
- · 5th
- * Deep Sky Objects

April mid-evening sky (March, late evening; May, dusk).

Eye, greatly outshining the modestly bright stars of the Hyades. The Hyades themselves are brighter than their neighbors, the Pleiades, but are less dramatic as a cluster because they are not concentrated into as compact a formation of glory as are the Pleiades.

If you look far up above Aldebaran and Orion in the middle of a winter evening, you will see a star considerably brighter than Aldebaran, one that even ever-so-slightly exceeds Rigel in brightness. This is palely yellowish Capella. It appears only one-third as bright as Sirius, but Sirius is the only star that outshines it on a winter evening. Capella is far enough north to have a second season of major visibility for most of the world's people. We'll discuss that when we get to autumn at the end of this chapter.

But we're not yet done meeting the 1st-magnitude stars of winter. In December, a very bright star rises in the mid-evening just before Sirius appears. Much later on a December night or, if you look a few hours after nightfall in February, this star shines only a moderate distance to the upper left of Sirius. Its name is Procyon, and it ranks after Rigel but usually before Betelgeuse in brightness. The constellation of Sirius is Canis Major, the Big Dog, and Sirius is therefore often called the Dog Star. The constellation of Procyon is Canis Minor, the Little Dog, so Procyon is sometimes called the little or lesser Dog Star.

Not too far above Procyon, we encounter not one but two very bright stars, separated by only slightly more sky than is filled by the compact line of Orion's Belt. The brighter of the two stars is Pollux; the dimmer is Castor. We mentioned them in the previous chapter's discussion of the dividing line between the stars of the 1st magnitude and those of the 2nd magnitude. Pollux is a 1st-magnitude star, but Castor is just a bit too dim and is therefore classed as one of the very brightest 2nd-magnitude stars. The two mark the heads of the famous twins of Greek mythology for whom they are named. And their constellation is indeed Gemini, the Twins.

The 1st-Magnitude Stars of Spring

Let's continue with our journey through the year, our trip to meet all the 1st-magnitude stars. In March and April, we see almost all of the bright stars of the Orion group setting sooner and sooner after the Sun, until each is lost in the solar afterglow. Not until late summer—and then only for people who look in the hour before dawn—does the "winter" host rise again. But if on spring evenings we look south and east, we can behold the fresh constellations of spring and, within them, the three beautiful 1st-magnitude stars of spring.

The first of these in its journey across the sky is the one we alluded to in the previous chapter as being the least bright of the 1st-magnitude stars: Regulus. Regulus first rises in the evening, a little north of due east, when winter is still young and Orion is not yet at his highest in the south. Likewise, on the first official day of spring, around 10:00 P.M., when Orion is not yet down to the western horizon, Regulus already stands at its highest in the south. So we can consider Regulus a star at its evening best in *early* spring. By the late nightfalls of June, Regulus is low in the west.

Although Regulus is the least bright of the 1st-magnitude stars, it marks the heart of great Leo, the Lion. Not only is Leo a prominent constellation, it is one of the constellations of the famous **zodiac**, which we'll discuss in the next chapter. It is noteworthy here to mention that the constellations of the zodiac are the ones visited by the moving Moon and planets. Regulus, in particular, is the 1st-magnitude star that is visited closely by the Moon and planets more often than any other. In addition, Regulus not only marks the heart of Leo but also the base of the handle of the big hook-shaped star pattern called "the Sickle."

A spring star much brighter than Regulus is pointed to by the curve of the Big Dipper's handle. Late on winter evenings, when the Big Dipper is getting pretty high in the northeast, this star peeks above the east-northeast horizon. The Big Dipper may have a striking form, but the six brightest of its seven stars are only 2nd-magnitude objects. The 1st-magnitude star the handle points to is the second brightest star visible from midnorthern latitudes, slightly outshining even Capella and Rigel. Its name is Arcturus.

Arcturus glows with a magnitude of –0.05 and has a distinct hue that can be vaguely termed light orange, though some observers want to call it peach-colored or slightly pinkish. Scientists have found out that Arcturus is the most unusual of all 1st-magnitude stars in its motion through space. We'll discuss this later in the book, in the profile of this star. For now, it's enough to know that Arcturus decorates the rather dimly starred expanses of the spring heavens with its brightness and appealing color. Not until June does Arcturus reach its lofty peak height in the south around nightfall.

The third and last of the 1st-magnitude stars in the traditional constellations of spring is Spica. It rises later than Arcturus and shines much lower in the southeast and south as the two progress across the sky. Spica, like Regulus, is a slightly blue-white star, but, also like Regulus, you probably won't see the color without binoculars unless the hue is exaggerated by the proximity of a yellow Moon or planet. For, yes, again like Regulus, Spica is the bright star of a constellation of that pathway of the Moon and the planets, the zodiac. In this case, the constellation is Virgo, the Virgin. Unlike Leo, Virgo is not very bright as a whole and is so large and asymmetrically shaped that it is hard for beginners to identify. Spica plays a vital role in helping the novice skywatcher locate this constellation.

A Far-South Digression

As spring nears its end in the Northern Hemisphere, people who live in southern Florida and southernmost Texas get a special treat. Climbing a little above their southern horizon for a while in May and June are four 1st-magnitude stars. They shine about five fist-widths at arm's length below Spica (this location is below the horizon if you live above about 30° N latitude). The brighter pair are those to the left (eastward), in the constellation Centaurus—Alpha Centauri and Beta Centauri. To the right (westward) from Alpha and Beta Centauri is one of the most famous of all star patterns—the Southern Cross. And in the four-star pattern of Crux, the Southern Cross, the two 1st-magnitude stars are Alpha Crucis (sometimes called Acrux) and Beta Crucis (sometimes called Becrux or Mimosa). Acrux marks the bottom (south end) and Becrux the left arm (east end) of the Southern Cross.

These four most brilliant stars of Centaurus and Crux are quite low in the south—and therefore dimmed by the thicker atmosphere low in the sky—even for the southernmost observers in the United States. That is also true of the extremely bright stars Canopus (visible to Florida and south Texas well below Sirius in winter) and Achernar (a lonely star visible to those geographic regions in autumn). But the farther south you go on Earth, the higher these stars get. It is from well down in the Southern Hemisphere—Australia and New Zealand, Argentina and Chile, South Africa—that these six stars appear very high. In fact, if you are far enough south, these stars don't set at all. Instead, they make big circles around the South Pole point of the sky, circles that are never cut off by the southern horizon. Any star that is far enough north or south to never set is called **circumpolar**. I've invented a special title for Alpha and Beta Centauri, Alpha and Beta Crucis, Canopus and Achernar: I call them the "South Circumpolar Six."

Folks in the southernmost United States see Alpha Centauri, Beta Centauri, Alpha Crucis, and Beta Crucis at their highest (not very high) on evenings in late spring and early summer. But seasons are opposite in the Southern Hemisphere. So in Australia, for instance, the Southern Cross is highest in the evening in late fall and early winter (May and June). Likewise, Canopus is highest in the evening during the Southern Hemisphere's summer (February); and Achernar, during the Southern Hemisphere's spring (November).

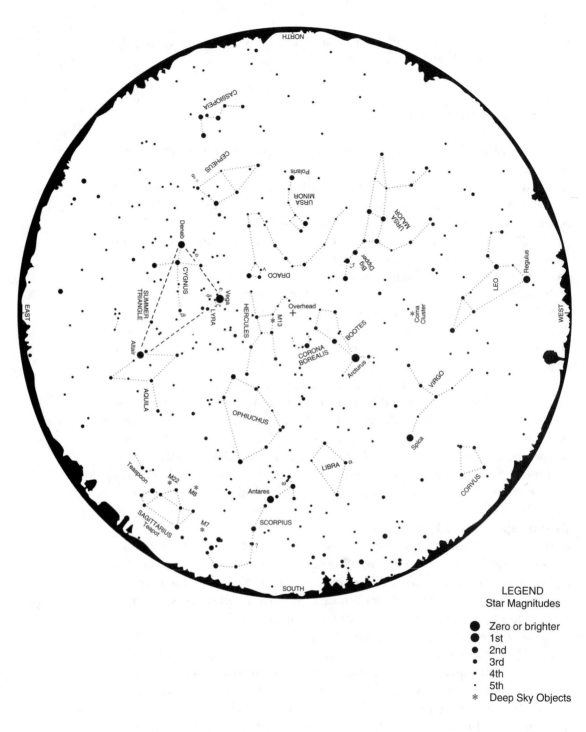

July mid-evening sky (June, late evening; August, early evening).

The 1st-Magnitude Stars of Summer

As May gives way to June, people at midnorthern latitudes are getting their longest days. But at nightfall, they see rising in the east an array of many more bright constellations and stars than spring had to offer.

At the end of a swelteringly hot day of glaring sun, a summer dusk often comes as a welcome relief. It also comes with a blue-white star that passes more nearly overhead than any other very bright one for people around 40° N, the world's most populous latitude. This lovely star is Vega.

Vega can be seen low in the northeast at nightfalls throughout much of the spring. But for many of us, June is the month when Vega starts getting high enough in the east to be seen over trees and buildings even before evening twilight fades. Vega is the third-brightest star visible from midnorthern latitudes, but the second-brightest, ever-so-slightly-brighter Arcturus, is still high in the south when Vega is midway up the eastern heavens. It's unlikely you'll confuse the two, though. Not only is Arcturus tinged with orange-gold and Vega with blue, but Vega, unlike Arcturus, is part of a pattern of three conspicuous stars that help identify it.

Once Vega is high enough, you will notice an only somewhat less brilliant star well to its lower right. This is Altair, head star of Aquila, the Eagle. Almost as far to the lower left of Vega is a bright star noticeably less radiant than Altair—but still brighter than anything else in its own section of the heavens. The third star is Deneb, the brightest luminary of Cygnus, the Swan. Together, Vega, Altair, and Deneb are the three 1st-magnitude stars that form "the Summer Triangle."

The Summer Triangle can be taken in pretty easily in one naked-eye field of view, but it is a big formation. Its two longest sides are longer than the Big Dipper (which at this time, by the way, is dropping lower in the northwest). The Summer Triangle is almost split in two by the cross-shaped Cygnus, which extends into the Triangle from Deneb.

Vega is at the top of the Summer Triangle as the pattern makes its long, glorious ascent up the eastern side of the sky. For viewers around 40°N, Vega reaches the **zenith**—the overhead point of the sky—around 1:00 A.M. daylight saving time in early July, 11:00 P.M. in early August, and 9:00 P.M. in early September.

After Vega starts down the western sky, Deneb travels very high (a little north of the zenith), with Altair passing much lower and into the southwest sky. Vega is visible for virtually every dark hour of every summer night. More surprising is the fact that the Summer Triangle's descent down the western sky lasts throughout autumn, all the way to the end of the year. You can see Vega and Altair very low in the west (with Deneb a little higher in the

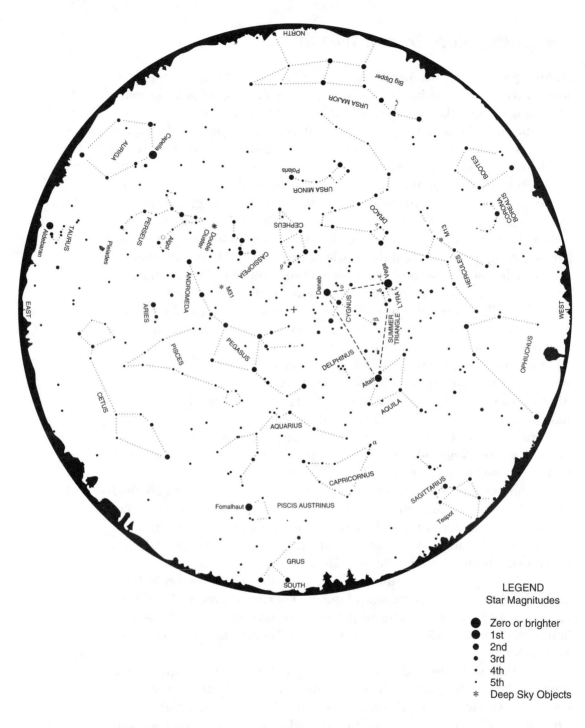

October mid-evening sky (September, late evening; November, early evening).

northwest) even in December, although you have to go out as soon as the sky gets dark.

The farther south a star is in the heavens, the lower and shorter the arc it will make across the southern sky for observers at midnorthern latitudes. The final very bright star of the traditional summer constellations is a southerly and therefore low star for observers in the United States and southern Canada. In this respect, it is the opposite of Vega. But this 1st-magnitude star is so distinctive that it deserves to share with Vega the title of quintessential star of summer. Its name is Antares.

Antares is one of only two very bright stars that shine with an almost ruddy—let's say golden orange or tiger-colored—light. The other is winter's great star Betelgeuse, which is even brighter than Antares. Like Betelgeuse, Antares is an enormous sun of the type called a red giant, and it varies in brightness (though less markedly than Betelgeuse does). The slightly ruddy hue of Antares is appropriate because it marks the heart of Scorpius, the Scorpion—summer's brightest and most striking constellation. The only problem with Scorpius and Antares, for many of us, is their southerliness and therefore lowness in the sky. Summer haze, always thicker low in the sky, often greatly dims Antares and the coiling twist of the other stars that make up Scorpius. On very clear (and Moon-free) evenings, after the passage of a cold front in summer, Antares and the Scorpion veritably blaze with prominence.

Such nights are also the time for observers far from city lights to get good views of a softly glowing band of radiance that is brightest from Cygnus down to Scorpius and its neighbor constellation Sagittarius. I'm talking about the Milky Way band—and will have more to say about it in upcoming chapters.

Antares is highest—though not very high—in the middle of July evenings. By September, we are getting our last good look at it, ever lower in the southwest at nightfall. We should be glad that the Summer Triangle lingers so nicely through autumn because of a remarkable fact: the traditional constellations of autumn, rising in the northeast, east, and southeast at nightfalls in September, contain only one 1st-magnitude star.

The 1st-Magnitude Star of Autumn— and Winter's Return

The solitary very bright star of the autumn constellations is Fomalhaut (pronounced *FOHM-uh-lawt*, not *FOHM-uh-low*). It is less bright than Antares (which itself is about midway in brightness between Altair and Deneb). It is also slightly farther south and thus even a little lower in Northern

Hemisphere skies than Antares is. Nevertheless, Fomalhaut is conspicuous on autumn evenings. The reason is that it shines in the midst of a vast region of unusually dim constellations that then fill the lower parts of sky from east to southeast to south. Fomalhaut stands out by itself like a lonely beacon.

Fomalhaut does get some help with brightening September, October, and November nights. The first source of help is a tower (or pile) of bright constellations rising from the northeast into the north sky. None of the stars in these constellations are 1st magnitude, but many are 2nd magnitude and form compact and interesting patterns. The second source of help is early sightings of the bright stars traditionally associated with winter.

The first of winter's stars to show itself again in the evening sky rises below the bright autumn constellations in the northeast: it is our old friend Capella.

Why would one of the winter stars, Capella, rise at about the same time as the autumn star Fomalhaut—as it does for viewers around 40° N? The northerliness of Capella is the reason. Whereas Fomalhaut is the southernmost of 1st-magnitude stars visible from around that latitude, Capella is the northernmost (slightly farther north than Deneb). From midnorthern latitudes, a star like Capella takes a long, high trip across the sky. But Fomalhaut follows a short, brief arc from southeast horizon to southwest horizon. The two can therefore rise together but, by the time Capella has finally reached its highest—at its halfway point in the trip across the sky, Fomalhaut is already setting in the southwest.

At around 40° N, Fomalhaut and Capella rise around nightfall in mid-September. But if Capella comes, can other stars of winter be far behind? Yes and no. Aldebaran doesn't rise at 6:00 P.M. standard time until around mid-November. But it comes up, straight below the Pleiades, around 8:00 P.M. daylight saving time in early October. A little after 8:00 P.M. standard time in mid-November, Rigel and Betelgeuse rise in the east, the three-star Belt is vertical between them, and at last one of astronomy's most awesome moments comes: we see the entirety of Orion's giant form tilted up above the horizon.

When Orion is rising, so too are Pollux and Castor, with the entire constellation Gemini. In less than two more hours, Procyon clears the horizon and, right after it, mighty Sirius lifts into view. That brightest of stars doesn't rise as early as 8:00 P.M. until mid-December. Winter is coming, the year ending. And a whole new year of bright stars is about to begin for us.

THE LOCATIONS, YEARLY MOTIONS, AND NAMES OF THE STARS

In this chapter, it is time for some explanation. We will first define exactly what constellations are. Also, having already encountered the stars in their seasons, we'll discuss why different stars and constellations appear in different seasons, marching onward in a stately progression. Even more important, we will see how astronomers measure distances and positions in the sky and how they fit all the constellations, north and south, east and west, onto the **celestial sphere**. Then, in the final section of this chapter, we will see, after stars are located, how astronomers have chosen to name them—often on the basis of the stars' locations.

Defining Constellations and Asterisms

The most famous and popular—and time-honored—means for locating stars is putting them in constellations. The word *constellation* literally means "a togetherness of stars." Throughout history (and even before), people would see patterns in the stars and trace them out star to star. The lines people traced helped form the figures—or at least the outlines—of things like animals, heroes, and monsters. These patterns became known as the constellations.

Eventually, in modern times, a more technical definition of *constellation* came into use. Astronomers decided not just to recognize patterns of stars but to consider a constellation as all of an area of the heavens containing (and, to a certain extent, surrounding) a pattern. Thus the heavens were apportioned into eighty-eight official constellations, official patches of sky that fit together like the pieces of a jigsaw puzzle. The drawing of boundaries for about half the constellations was accomplished by Benjamin Gould in 1875. The job was completed by Eugene Delporte and approved by the IAU (International Astronomical Union) in 1930.

The modern system does not grant constellation status to some famous star patterns that tradition considered parts of larger constellations. The seven prominent stars of the Big Dipper (known as the Plough in Great Britain) are regarded as part of an even huger pattern—Ursa Major, the

Great Bear. Ursa Major is the official constellation. The Big Dipper is an **asterism**—an unofficial star pattern. There are several very big asterisms that stretch across the borders of some of the official constellations, borrowing stars from them. A famous example is an asterism mentioned in the previous chapter: the Summer Triangle. The Summer Triangle is composed of Vega (from the constellation Lyra), Altair (from the constellation Aquila), and Deneb (from the constellation Cygnus).

About half of the official constellations are ones that have been known since ancient times. The others are later inventions, dating mostly from the sixteenth, seventeenth, and eighteenth century. Most of these latter-day creations were made to organize the stars of the far south regions of the heavens, which had never been glimpsed from ancient Rome and Greece or from medieval Europe. Astronomers gave many of the new constellations names like Horologium, the Clock, and Microscopium, the Microscope, reflecting their interest in modern, sometimes scientific devices.

The Seasonal Progression of Constellations

In the previous chapter, we noted that the special constellations of each season progress across the sky earlier and earlier each night until they start setting too soon after the Sun to see. The winter constellations, we found out, start setting with the Sun in spring and reappear just before dawn in late summer. Then they keep rising longer and longer before sunrise until, by autumn's end, even the last of the winter stars and constellations are rising before midnight.

What is it that causes this seasonal progression of the constellations? The explanation is simple. The stars are so far away that they are virtually standing still in relation to our vast distance from them. What is actually moving close enough to us to make the distant stars seem to be moving during the course of the year is Earth in its orbit. After all, the year *is* Earth's making one circuit of its orbit around the Sun.

Here's how it works. As the year and its seasons progress, Earth's motion in its orbit makes each season's constellations get closer and closer to our line of sight with the Sun. What that means is that we see those constellations set sooner and sooner after the Sun does. Then come about two months when those constellations are approximately on our line of sight to the Sun—actually, on the far side of the Sun, with the solar glare between us and them. Then Earth moves on and we can see these constellations in question start to appear farther and farther to the other side of our line of sight to the Sun—in other words, they keep rising longer and longer before the Sun does. Eventually, the constellations are again in line with Earth and the Sun—but this time in the direction opposite from the Sun. So we then see the constellations rise around sunset, be at their highest in the middle of the night, and set around sunrise. What's the final stage? The star patterns are again setting sooner and sooner after the Sun does, their prime season of visibility beginning to pass. And so the cycle begins all over again.

How much seasonal progression of the constellations occurs from one night to the next? This is an interesting and important question. The answer is about four minutes. In other words, stars rise (and set) about four minutes earlier each night.

When I wrote previously that the Sun is "between" us and a group of constellations, I said approximately not precisely. This is true for most of the constellations. The Sun might be somewhat north or south of a constellation—but of course the solar glare extends so far that it would still make the constellation unviewable. There are, however, certain constellations that the Sun does truly pass directly, exactly in front of, from our point of view on Earth. We can therefore say that the Sun is "in" these constellations (though their stars are actually tremendously far behind the Sun in space, of course). During the course of a year, the Sun passes through a series of these constellations and slowly goes around an entire circle of the heavens (a reflection of the circle of Earth's orbit). This series—the constellations the Sun seems to travel through during the course of a year—is called the zodiac.

The Zodiac Constellations, and Conjunctions

A zoo is a place where animals are kept; zoology is the study of animals. The word *zodiac* is from the ancient Greek term *zodiakos kuklos*—the "cycle of little animals." The animals in question are, of course, the constellations. The ancient Greeks noticed that all the constellations of this cycle, or circle, did more or less represent animals, for instance, Taurus, the Bull, and Leo, the Lion. I say "more or less" because this statement is only completely true if you count such human figures as Gemini, the Twins, as technically animal—and (for this purpose only) discount Libra, the Scales. But there is reason to do the latter: Libra, representing the Roman scales of justice, was really only created in late classical times, from what had previously been the claws of Scorpius, the Scorpion. The original Greek conception of the zodiac did not include Libra.

There are now commonly held to be twelve constellations of the zodiac—

one for the Sun to be in each month of the year. But if we examine the modern boundaries of the constellations, we find that the Sun spends time in thirteen, not twelve. The extra constellation, which nonastronomers have not heard of, is Ophiuchus, the Serpent-Bearer. Amazingly, the Sun actually spends a lot more time within the modern boundaries of Ophiuchus than it does in Scorpius, one of the famous ancient members of the zodiac.

The twelve traditional constellations of the zodiac that everyone has heard of are, in their customary order: Aries, the Ram; Taurus, the Bull; Gemini, the Twins; Cancer, the Crab; Leo, the Lion; Virgo, the Virgin; Libra, the Scales; Scorpius, the Scorpion; Capricornus, the Sea-Goat; Aquarius, the Water-Bearer; and Pisces, the Fish. If they all sound familiar to you, it is partly because these constellations have also been called "the signs of the zodiac" and are thought by astrologers to influence the lives of people when the Sun, the Moon, or certain planets are within their boundaries. There is, in fact, no scientific evidence that astrology is true nor even a reasonable explanation of how it could be true. But the constellations of the zodiac are interesting for astronomers to observe—especially because of the visiting solar system bodies we can see among their stars.

Yes, not just the Sun but the Moon and planets can always be found in the zodiac. The Sun's precise path through these constellations, what we might consider the midline of the zodiac, is called the **ecliptic**. It is really a projection of Earth's orbit in space (again, it is Earth that does the moving, not the Sun). The orbits of the other planets (and the Moon's) lie in nearly the same plane as that of Earth's orbit, so we never see these worlds wander very far from the ecliptic. The 1st-magnitude stars closest to the ecliptic are therefore frequently visited by the Moon and planets.

When a solar system body such as the Moon or a planet passes to the north or south of such a star (or north or south of each other), we call that event a **conjunction**. (This is the strict definition. In a looser sense, a conjunction is merely any close temporary pairing of celestial objects.) On some occasions, the Moon may actually pass not north or south of, but directly in front of, a planet or a bright star, an event called an **occultation** (the "occult" is supposed to be the hidden and these events are a hiding of one celestial object by another—although, in some cases in astronomy, a hiding is called an **eclipse**).

The four 1st-magnitude stars that can be occulted by the Moon and that often have close conjunctions with planets are Regulus (in Leo), Spica (in Virgo), Antares (in Scorpius), and Aldebaran (in Taurus). Pollux (in Gemini) is a little too far from the ecliptic to be occulted by the Moon but sometimes has rather close conjunctions with the Moon.

Positions in the Sky: Altitude and Azimuth

When we talk about stars (or other **celestial objects**) we need to have a system for identifying where they are in the sky. Of course, we can say that the star Betelgeuse is in the constellation Orion. But to specify exactly where Betelgeuse is and be able to compare readily where it is with the location of any other celestial object or region requires a system that involves numerical values.

The simplest such scheme is called the **altazimuth system**. It tells us where an object is in the hemisphere above the land and sea that we call the sky. **Azimuth** is the sideways measure in the sky. **Altitude** is the up–down measure in the sky. Don't confuse this kind of altitude—which is angular altitude given in degrees (as in 360° in a circle)—with altitude in Earth's atmosphere—which is given in feet or meters or miles or kilometers above Earth's surface. Azimuth is also angular, also given in degrees.

Angular altitude is perhaps a little simpler than azimuth. Two fundamental positions serve as limits at the extremes of altitude: the **horizon** (the line that forms the border between sky and land or sea) and the **zenith** (the point exactly overhead). The horizon is set at 0°. If we draw a line from one horizon point straight up through the zenith to the opposite horizon, this distance would be 180°—a dome or half-sphere that is one-half of the full 360° circle that passes both above our heads and below our feet. And since the distance from horizon to zenith alone is half of 180°—90°—we say that the altitude at the zenith is 90°. Thus, in the altazimuth system, a rising star that is right on the horizon would be at an altitude of 0°; a star that is exactly overhead would be at an altitude of 90°. Any angular height in the sky between these extremes would be a value in degrees between 0° and 90°, of course.

Azimuth is the sideways frame of reference in the sky—that is to say, where an object is in relation to the cardinal directions of north, east, south, and west. That is, by the way, the progression of directions used and, to numerize them, we set due north as equal to 0° azimuth, east as 90°, south as 180°, west as 270°, and, to complete the circle around the sky, north is 360°—or, which is the same thing, 0° again. Thus, instead of having to say that the star Regulus is currently a little south of south-southeast, we could say with much greater precision and economy of expression that the current azimuth of Regulus is 165°.

The altitude and azimuth of all celestial objects must be given for a specific time because these figures keep changing as Earth carries us around with its rotation making stars and other celestial objects appear to move during the course of a night or day. The only object that wouldn't change altitude and azimuth would be one located exactly above (north of) or below (south of) Earth's axis of rotation. The bright star—one of 2nd magnitude—which comes closest to meeting this condition is a famous one: it is Polaris, the North Star.

The mention of Polaris brings us to the other most important—in fact, much more important—positional system used by astronomers: declination and right ascension.

Positions in the Heavens: Declination and Right Ascension

The other system used by astronomers is called the **equatorial system**, and it deals with the heavens, not the sky.

The *sky* is the infinitely distant dome of what may be seen everywhere above the horizon. There are many skies, depending on one's location on Earth. The *heavens* are the infinitely distant sphere in which are located all celestial (which is to say, heavenly) objects both above and below an individual observer's horizon. In short, the heavens that comprise all of our skies encircle the entire Earth. Altogether, they form what astronomers call the *celestial sphere.*

Picture the sphere of Earth inside this celestial sphere. Every position of latitude and longitude on Earth can be said to have its counterpart above on the imagined inside surface of the celestial sphere. But on the celestial sphere, in the equatorial system, *declination* corresponds to latitude and *right ascension* to longitude.

Declination is the north–south frame of reference on the celestial sphere. The spot on the sphere directly over Earth's North Pole is called the *north celestial pole* and over Earth's South Pole is called the *south celestial pole.* Halfway between the poles, hanging directly over Earth's equator, is the *celestial equator.* As on Earth, the celestial poles are 90° from the celestial equator, which is the 0° line of declination. But whereas we speak of 90° north latitude (or 40° north or whatever), we speak of +90° declination. Likewise, the south celestial pole has a declination of –90°.

Right ascension—usually called RA for short—is the east-west frame of reference on the celestial sphere. It must be measured from a prime meridian (initial line of RA) just as longitude has to be on Earth. On Earth, the prime meridian—0° longitude—was set to pass through the famous Greenwich Observatory in England. Longitudes are measured in degrees—0° through 180°—east or west of the prime meridian. Right ascensions, however, are measured just eastward all the way around the celestial sphere—in *hours* of RA (from 0 hours or "0h" through twenty-four hours past "23h" and

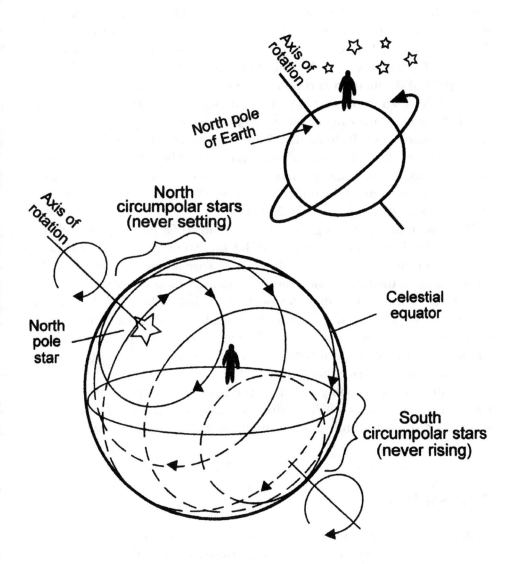

Axis of rotation

North pole
of Earth

North
circumpolar stars
(never setting)

Axis of rotation

North
pole
star

Celestial
equator

South
circumpolar stars
(never rising)

The celestial sphere. Upper right diagram: We stand on a world that is rotating eastward. Lower left diagram: The appearance, however, is that we stand on a motionless world within a celestial sphere that is rotating westward. As seen from middle latitudes in the Northern Hemisphere, stars near the celestial sphere's north pole never set and stars near the celestial sphere's south pole never rise (the latter are always hidden from the mid-northern latitude observer by the bulk of Earth).

onward through the final hour back to 0h). The prime meridian of the celestial sphere—the 0h line—has been decided to be the line of RA that passes through the "vernal equinox"—the position the Sun reaches at the opening moment of spring in Earth's Northern Hemisphere.

Degrees of declination and hours of right ascension are angles that are divided into smaller units for precise positions. One degree of declination equals 60 *minutes of arc* or *arc minutes*; 1 minute of arc equals 60 *seconds of arc* or *arc seconds*. For declination, minutes of arc are indicated by ' and seconds of arc by ". Thus the declination of a star might be –12° 20' 30" (in practice nowadays, one sees more often the seconds omitted in favor of using a decimal fraction form of the minutes—in this case, for instance, –12° 20.5'—30", remember, would be half, or .5, of 1').

Right ascension is already given in the large units of hours, so what could be more natural than dividing one hour of RA into 60 minutes and one minute into 60 seconds? Don't forget that these are still arc minutes and arc seconds, angular units in the heavens, not simply minutes and seconds of time.

It is true, however, that these terms of RA, including hours of RA, do also have an origin and meaning in time. The rotating Earth makes one hour of RA in the heavens move past any fixed element—for instance, a tree in our landscape or the altazimuth system direction called south—in one hour. I must hasten to add that there is a small deviation from this, due to the fact that Earth is not just rotating but also moving along its orbit around the Sun. I'm referring to the fact, which we discussed earlier, that the stars rise, set, or reach any given altazimuth position about four minutes earlier each successive night.

Universal Time

Under a sky that seems to be turning above the still Earth, we come to realize that time and position are intimately connected. What kind of time system do astronomers use to keep track of when celestial events occur? They use **Universal Time**, also known as UT.

Universal Time is basically the same as 24-hour (no A.M. or P.M.) standard local time on Earth's 0° of longitude—the meridian that passes through Greenwich in England. The Eastern Time Zone of North America is five hours behind Greenwich Time and therefore five hours behind UT—when daylight saving time is not in effect (when daylight saving time *is* in effect, Eastern Daylight Time is only four hours behind UT, which is never itself subjected to daylight saving time). Thus, on a date when standard time is being used, 10:30 UT would be equivalent to 10:30 minus 5, which equals

5:30 A.M. Eastern Standard Time. But 10:30 UT would be 6:30 A.M. EDT. Since UT is a 24-hour system, 19:00 UT is 19 minus 5, which equals 14—or, rather, 2 hours after 12 noon, which equals 2:00 P.M. EST.

Naturally, if you don't live in the Eastern Time Zone, you have to subtract or add a different amount of time to UT. In North America, Central Standard Time is 6 hours earlier than UT, Mountain Standard Time is 7 hours earlier than UT, and Pacific Standard Time is 8 hours earlier than UT.

There's a final twist for many of us: 0 UT is the start of a Universal Time day—and also the start of the day, midnight, on the Greenwich meridian. But in the Eastern Time Zone of North America, the time is 5 hours standard time earlier—which is 7:00 P.M. *on the previous calendar date.* Thus, 3 UT on February 15 is 10:00 P.M. Eastern Standard Time on February 14.

Measuring Sky Angles with Your Hand

There are degrees of declination and hours of right ascension. But suppose we want to measure the length of a star pattern or comet tail that is not oriented exactly north–south or east–west. We can measure the angular distance between any two places in the sky in terms of its fraction of 360°—and with an instrument as simple and convenient as our hand. The width across a person's own fist as viewed by him or her at arm's length is about 10°. This is approximately true whether you are a child or a tall basketball player, because short people with short arms tend to have small hands whereas tall people (with long arms) tend to have large hands. We know that the distance from the horizon straight up to the zenith is 90° (half of the way across the 180° half-sphere of the sky), so try measuring it with your fist at arm's length. You should be able to fit about nine fists—9 × 10°= 90°.

You'll find that the Big Dipper's length is about 2½ fist-widths at arm's length—approximately 25°. The distance between the stars on the top (widest part) of the imagined bowl of the Big Dipper is 1 fist-width—10°. The distance between the Pointers (two stars on the side of the bowl opposite from the handle, which point to the North Star) is about half a fist-width—5°. Most people's little finger is about 1½° wide when held at arm's length. This can be useful to estimate the distance between a planet and a star when they are in conjunction.

Some conjunctions are closer (less separation between objects) than 1½°. For these and other quite small angular distances or diameters in the sky or heavens, we can calculate the diameters of the narrow fields of view we get in a telescope by using different eyepieces and comparing them to a conjunction separation or apparent width of a planet.

When we consider such celestial objects as planets or galaxies, we need tiny units of angular measure. So, once again, we can use arc minutes (') and arc seconds ("). The Moon and the Sun are both about $1/2°$ wide—or about 30'. The planet Jupiter always appears less than 1' (less than 60") wide—in fact, never any more than 50" in diameter. The smallest angular distances commonly used in amateur astronomy are the separations between the two very close-together stars in a "double star." These separations can be less than 1" and are visible only with medium to large telescopes on nights when our atmosphere is quite steady and star images are sharp. All these stars are suns—some, giant suns—but they are so far away that no telescope can directly show them as sizable objects. But atmospheric turbulence and the limitations of telescope optics can enlarge the image of the star to usually somewhere between 0.5" to 3", depending on the night and the telescope.

Circumpolar Stars

Now let's expand our vision from the angularly very small to the very large. Now that we know about RA and declination and especially about the celestial poles and celestial equator, we can better understand how much of an observer's sky is filled with **circumpolar** stars.

Imagine yourself at Earth's North Pole. What do you see overhead? The north celestial pole with the North Star, Polaris, very near it. If Polaris was exactly at the north celestial pole, it would never change its position even in the slightest. Instead, Polaris makes a tiny circle around the north celestial pole (this is true whether we are at Earth's North Pole or somewhere else on our planet). But what do the stars in the rest of the sky do? As seen from Earth's North Pole, they all circle around the sky, maintaining the same altitude, with the north celestial pole as their center. In other words, none of the stars ever set: all are circumpolar. From this unusual spot on Earth, the North Pole, the celestial equator lies along the horizon. Here at the top of Earth's Northern Hemisphere, the sky is occupied by the north celestial hemisphere—and all stars south of the celestial equator are forever hidden below the horizon.

Now imagine yourself at Earth's equator. Here, the celestial equator runs from due east to overhead down to due west. Where are the north celestial pole and the North Star? They are right on the due north horizon (if you can even glimpse Polaris through horizon mist and the longer path of air down low in the sky). What is on the opposite—the due south—horizon? The south celestial pole—which doesn't happen to have any prominent star near it. How many stars in this sky at Earth's equator are circumpolar, never setting? None. On the other hand, we can at one time or other see all stars from there. The Big Dipper, which is never seen by observers south of a certain latitude in Earth's Southern Hemisphere, and the Southern Cross, which is never seen north of a certain latitude in Earth's Northern Hemisphere, can both be seen from Earth's equator (though only part of the time).

The vast majority of us, of course, live somewhere between Earth's equator and either the North Pole or the South Pole. The most highly populated latitude is around 40° N. Let's choose it as an example of a latitude between the two extremes of the equator (0° latitude) and a pole (90° N or 90° S latitude). We are in Earth's Northern Hemisphere, so the North Star must be visible. Where is it? It is in the north sky (of course!), but how high above the north horizon? The answer is 40° (four fist-widths at arm's length). This interesting relation holds for any latitude—whatever the numerical figure for your latitude is, that is the angular height of the North Star (or, more precisely, the north celestial pole). The farther north you go, the higher in the north Polaris gets. The farther south you go, the lower in the north it gets. (Of course, when you go south of the equator, Polaris goes below the north horizon and is blocked from your view by the curving bulk of Earth. But even there, the mathematical relation works if you use negative numbers—at 20° S latitude, which could be considered –20° latitude, Polaris will be at an altitude of –20°, which means 20° below your north horizon.)

Now, if we return to our site at 40° N and find Polaris 40° high in the north, how many stars will be circumpolar—that is, close enough to the North Star in the heavens to never have their circles around it cut off by an east or west horizon? Well, a star passing under the North Star at 40° N latitude would have to be no more than 40° from the North Star to avoid having to dip below the north horizon for a while. Since the declination of Polaris—or, rather, of the north celestial pole—is +90°, a star must have a declination of greater than (farther north than) +90° minus +40° equals +50° declination to be a north circumpolar object. Only one star in the Big Dipper—the star at the end of the handle—has a declination of less than +50°. That star's declination is just over 49° so, for viewers at 40° N, it ever so slightly dips below the due north horizon when it is directly under Polaris.

The farthest north of the 1st-magnitude stars is Capella. Its declination is exactly 46°. So anywhere north of 44° N latitude (90° minus 46° equals 44°), Capella will never set, even at the due north point of the horizon.

We can also use our latitude to figure out the declination of stars that pass overhead. At 40° N latitude, the north celestial pole is 40° above the north horizon. Traveling another 50° straight up to reach the zenith (its altitude is 90° and 40 + 50 = 90), we find that if a star at the zenith must be 50° farther south than the north celestial pole, then its declination must be 90 minus 50, which equals +40° declination. That's again (as with the angular height of the North Star) the same number as our latitude.

How do we use our latitude to figure out the farthest south declination a star can be to ever appear above our south horizon? We take the declination of the zenith and then subtract 90° from it. So from 40° N latitude, where stars at the zenith are at +40° declination, the farthest south star visible is +40 minus 90, which equals –50° declination.

Precession and the Succession of North Stars

Now if the north and south ends of Earth's rotation axis always pointed in the same directions, we could stop here in our discussion. But Earth undergoes a wobble similar to that of a spinning top before it falls. Earth is not going to stop revolving and fall over on its side, though. In this case, the wobble is not caused by a loss of momentum but by a pulling on the rotating Earth by the Sun, the Moon, and planets. These bodies don't change the amount of tilt the rotation axis has to the plane of Earth's orbit—that remains at $23^1/_2°$. But the direction the axis points falls around the north pole of Earth's orbit (the ecliptic pole), tracing out a cone—or rather two cones, one to north and one to south. The radius of the cone is $23^1/_2°$. And that is the radius of the circle that the north celestial pole slowly makes against the background heavens. How slowly? That is the remarkable thing. This wobble of Earth is called **precession**. And one cycle of precession—one trip around the $23^1/_2°$ radius circle for the north celestial (and south celestial) pole—takes 25,800 years!

Of course, that is a very long period of time by human standards. You might guess that people could scarcely notice the effects of precession in the course of recorded human history, let alone in a human lifetime. But some signs of precession in the heavens are more rapidly noticeable—and important—than you might think.

First, there is the change in which a star is near enough to the north celestial pole to be called "the North Star." This doesn't change radically in a few hundred years, but it does in a somewhat longer period. Right now, we are about one century away from the north celestial pole's closest pass of the star we currently call Polaris. But two thousand years ago, two other stars in Ursa Minor (the Little Bear, whose pattern is also called the Little Dipper) earned their title of "Guardians of the Pole" by being near it. And over four thousand years ago, when the Egyptians were already a mighty culture, the North Star was in another constellation altogether—it was the star Thuban, the

Alpha star of Draco, the Dragon. Go back to about 12,000 BC, around the time of the last Ice Age, and the north celestial pole was roughly halfway around its full circle of precession and was actually rather close to the magnitude 0.0 star Vega (remember, Polaris is only magnitude 2).

But precession causes more than just the north celestial pole to change position. It is slowly shifting the constellation the Sun is in at the times of the solstices and equinoxes. The "precession of the equinoxes" has moved the vernal equinox from Aries into Pisces in the past two thousand years and in a few hundred more years it will move west into Aquarius. That time will truly be the beginning of the "age of Aquarius." However, astrology has stuck with the "houses" or "signs" of the zodiac as they were two thousand years ago. Thus, when traditional astrology says that your sign is Leo—meaning that you were born when the Sun was in Leo—the Sun was actually not in Leo but one to two constellations west of it.

So precession changes the position of the celestial poles and the equinoxes and solstice points of the Sun among the constellations. But if the celestial pole moves in relation to the stars, then the entire grid of right ascension and declination moves with it. Unfortunately, when astronomers try to find a faint object at a precise position, that location may have changed not in a thousand or a hundred years but in just a few years (or, in the most extreme cases, even a few days). For most purposes, RA and declination will be sufficiently precise if they refer to an epoch that lasts for fifty years. In astronomy books up until about 1975, you will see RA and declinations listed for epoch 1950.0 (the start of the year 1950). After that, the epoch given is 2000.0—and will be until about 2025, when observers will start using epoch 2050.0.

Star Names and Designations

In this chapter, we've defined what constellations are and then studied two common positional systems used to locate the stars. It is additionally essential to establish each star's identity by a system for naming or designating it.

Stars have been given proper names throughout history. Names like Sirius, Regulus, Aldebaran, Betelgeuse, and others serve well to identify the bright stars they are attached to. But as astronomers extended their studies down to fainter and fainter stars, it became necessary to distinguish among the ever greater numbers of stars by a new means: catalog designations.

Four hundred years ago, the astronomer Johannes Bayer helped establish the Greek letter system for designating stars. In this scheme, still in use, the few dozen brightest stars in a constellation could be given letters of the

Greek alphabet in combination with the genitive forms of the constellation. Thus the brightest star in Leo could be called Alpha Leonis (although it also has the proper name Regulus). The rule ought to have been to apply *alpha*, first letter of the Greek alphabet, to the brightest star in a constellation; *beta*, the second letter, to the second-brightest star; and so on down through the alphabet in order of decreasing brightness. Unfortunately, in some constellations, the star called Alpha was only the second brightest or was even fainter relative to other stars in the configuration. Furthermore, the letters were sometimes assigned according to another rationale (or lack of rationale) that no one now can figure out. In several important constellations, the guiding principle was not order of brightness but is at least comprehensible; for instance, in some constellations the letters were assigned to the stars of a prominent pattern in an order working positionally around the pattern. The best example of the positional approach is probably the Big Dipper part of Ursa Major. There, the first seven letters of the Greek alphabet have been assigned to its stars in order from the lip of the bowl around the bottom and out to the end of the handle.

In 1783, the French astronomer J. J. de Lalande reproduced an earlier star catalog created by the British astronomer John Flamsteed. Lalande applied numbers to the stars in the catalog in order of increasing RA (that is, west to east) within their constellations. These became known as "Flamsteed numbers." An example of one of these designations would be "118 Tauri"—the 118th star, in order of increasing RA, in the constellation Taurus.

But even the Flamsteed catalog covered only stars visible down to around the naked-eye limit. Later star catalogs have grown ever more extensive, eventually becoming little more than statements of the RA and declination of stars—really just addresses, not names.

Fortunately, all but a few of the 1st-magnitude stars have proper names—names usually both time honored and rich with lore. A few of the very bright stars of the far south, essentially unknown to European civilization until the sixteenth- and seventeenth-century ship voyages of discovery, are usually called by their Greek letter designation. Even with these, there have been modern attempts to give them proper names. "Hadar" has been applied to Beta Centauri, "Mimosa" to Beta Crucis—and the half-baked names "Rigil Kentaurus" and "Acrux" for Alpha Centauri and Alpha Crucis, respectively.

Seeing Stars Better (Skies, Eyes, and Telescopes)

4

Learning how astronomers measure the positions of stars and understanding in what fashion stars move during the course of a year is very important. We can't observe particular stars if we don't know where to look for them.

We had quite a lot to say on these topics in the previous chapter. But what happens once we do figure out the positions of the stars we want? How well we may actually see them will depend strongly on how much light pollution we have at our observing site, how much haze is in the sky, how turbulent the atmosphere is, and how bright the Moon is if it is in the sky. Seeing stars optimally depends also not just on the quality of our skies but on how we use our eyes. Simple viewing techniques can make a world of difference in what we see of stars and other dim celestial objects. Finally, we all know that there are instruments—called binoculars and telescopes—that can increase the natural powers of human vision. What are the absolute essentials we need to know about these devices? This chapter takes a quick look at all these topics.

Light Pollution

We begin with the greatest impediment of all to observing the heavens in today's world. **Light pollution** is excessive or misdirected artificial outdoor lighting. It does no one any good, and almost everyone considerable harm. Glare in the eyes of motorists and homeowners greatly reduces traffic safety and interferes with crime prevention. Properly shielding lights can reduce or eliminate these problems. Light pollution also has adverse, sometimes fatal, effects on many birds, animals, and plants. Excessive lighting at night has even been found to reduce production of melatonin in the human body and lead to increased incidence of certain cancers, including breast cancer. Again, proper shielding of lights and otherwise judicious use of them can mitigate these problems.

One key component of light pollution is both the costliest of all and the one that affects stargazers the most. **Skyglow** is light that escapes upward to illuminate the air above cities and other light-pollution sources. It not only

washes out city-dwellers' views of the stars but even as much as dozens of miles away from a large city it can ruin a section of sky for a rural skywatcher. Skyglow alone is costing the United States several billion dollars a year in wasted energy (millions of barrels of oil, millions of tons of coal). Yet skyglow can be greatly reduced by simply using "full-cutoff" shielded lights. Indeed, scientists estimate that about three-quarters of all existing light pollution can be eliminated by the use of already existing technology and practices. Doing so will not just save money, it will reduce the need for energy and thereby lessen the burning of fossil fuels and all the problems that come with them: air and water pollution, global warming, and the destruction of wildlife habitats that usually accompanies the building of power plants.

So how can you, an individual, help in the battle against light pollution? The first step is to become educated by reading up on the subject—preferably first at the Web site of the International Dark-Sky Association (IDA), www.darksky.org. IDA is a twenty-year-old organization that now has more than eleven thousand members from all over the world. You can act globally by joining IDA and locally by learning from IDA how to encourage local government and businesses to address the problem.

What can you do to get darker skies *tonight*? Not much, of course, except when you are able to drive to a darker site. You can look up light-pollution levels in your county, state, or province by going to the excellent Clear Sky Clock Web site, www.cleardarksky.com. That site is also where you can get predictions for the next two days of cloud, transparency, and "seeing" conditions in your area. Let's examine these factors.

Transparency and "Seeing"

All of us understand that cloudiness can block our view of the heavens. But it's also crucial to realize that haze, usually but not always associated with high humidity, is another important consideration. What we want is not only a sky free of clouds but a "transparent" sky.

Transparency is the measure of how much light can pass through the atmosphere. A night of bad transparency will show a few feeble stars in a sky that looks washed out or even whitish. It usually follows a day when the sky was more white than blue and there was a lot of water vapor (humidity) in the air. But transparency can be reduced also by dust, pollen, and especially sulfuric acid haze from the sulfur dioxide released in vast quantities by coal-fired plants. A night of good transparency has the darkest sky and brightest-looking stars—at least if you are away from light pollution and bright moonlight (although even these cause less brightening of the sky when the sky is free of aerosols and particulates).

What else could an observer ask for than the darkest, most transparent sky possible? Well, it often turns out that a transparent atmosphere is one that has been whisked clean by the recent passage of a cold front. But such fronts (as well as many other weather features) are usually associated with turbulent air (and not just at the surface— often the kind of turbulence that makes the stars twinkle more than usual occurs higher in our atmosphere). The hard fact is this: turbulence leads to shaky, fuzzy images of astronomical objects.

Seeing is the degree of sharpness of images as a function of how steady the atmosphere is. Seeing is especially important for lunar and planetary observation—for crisp, detailed images of things like Saturn's rings or a lunar crater. But it's also important for some aspects of stellar observation— for instance, our ability to cleanly separate the components of a tight double-star system. Seeing is sometimes best when the atmosphere in your region has been so calm (perhaps near the center of a high-pressure system) that the air mass actually becomes stagnant. Unfortunately, that is when haze and air pollution can then worsen and greatly reduce transparency. Nevertheless, it's not hopelessly difficult to get a night of both good transparency and good seeing. When an astronomer does, though, he or she should be truly elated.

The Phases and Hours of the Moon

Most beginners are surprised when they learn that amateur astronomers usually try to avoid having a bright Moon in the sky (when they are not observing the Moon itself, of course). What bothers astronomers is that strong moonlight brightens the sky a lot and makes all but the brightest other objects in the heavens far more difficult to see. There is a certain spartan beauty to when only the brightest stars are visible in bright moonlight and look like intense little specks. There's no denying, however, that the brightest stars look far more lustrous and vivid when the Moon is out of the sky or at least very thin.

Did you know that the Moon is sometimes below the horizon for part of a night—or even the entire night? What you need to learn, if you are to make the most of the night's moon-free hours of darkness, is how the where and when of the Moon is inextricably connected with its phases.

New Moon is when the Moon is passing due north or due south of the Sun (on rare occasions, centrally in front of the Sun for a stunning total eclipse of the Sun or interesting "annular" eclipse of the Sun). If there is not an actual eclipse showing off the Moon in silhouette against the Sun, the New Moon is too deep in the solar glare for us to see. In this phase, it is also exposing to us nothing but its dark, night side—so, for that reason, too, we can't see the Moon. But we do get to have Moon-free skies all night and have no interference from moonlight for our viewing of the stars.

The Moon does not stay still, however. Its eastward motion in its orbit makes it appear about 12° farther east at the same time on two succeeding nights. This means that the Moon rises—or sets—about one hour later each night. A few days after New Moon, this motion takes it far enough out to the side from our line of sight to the Sun for us to see the Moon, low in the west right after sunset. Only a narrow sliver of this Moon's west-pointing edge is sunlit. It is a waxing (enlarging) crescent. But it is not bright enough to wash out the entire sky, and it sets only a few hours after the Sun—after which the rest of the night is Moon-free for stargazing.

A week after New Moon, the Moon is half lit—the half toward the Sun, toward the west. But the Moon is only one-quarter of the way through its complete progression of phases. Therefore, we call this phase First Quarter Moon. It stands in the south at sunset and doesn't set until the middle of the night. Its light is bright enough to somewhat compromise observation of very dim stars and other objects. And it is only out of the sky in the hours between about midnight and dawn.

When the Moon is less than half lit but more than New Moon, we call it a *crescent*. When it is more than half lit but less than fully lit, we call its phase *gibbous*. The waxing gibbous Moon keeps appearing farther east at each nightfall—and it sets later and later in the night. It is bright enough to seriously compromise much stellar observation and only exits to leave the sky really dark at the final hour or so before dawn.

Finally, the Moon is rising right around sunset. It is then in the opposite position from the Sun and therefore pointing the same face toward us as it is toward the Sun—in other words, we see the entire daytime side of the

SEEING THE SOLAR SYSTEM

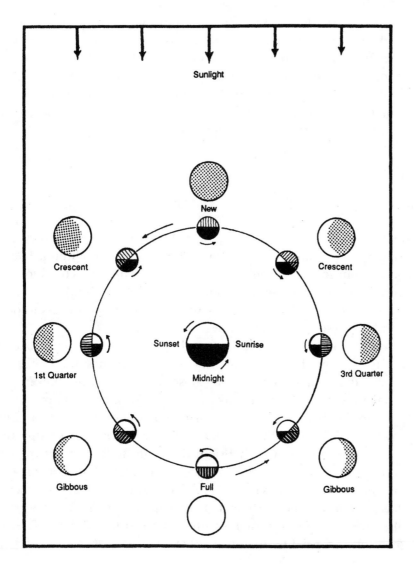

The cause of the phases of the Moon.

Moon. Now the Moon's earthward face appears fully lit: it is Full Moon. Full Moon is bright enough to spoil much stellar observation. Not only does it rise around sunset, it is highest around the middle of the night and doesn't set until around sunrise. Full Moon is in the sky all night.

But each night after Full Moon, the Moon rises an average of a little less than an hour later. Progressively, more and more of the evening is Moon-free and the Moon is getting less bright. This waning gibbous shrinks until, about one week after Full Moon, the Moon rises in the middle of the night, appears half lit (the east-pointing half), and is called Last Quarter Moon (for it is starting the final quarter of its run through the complete progression of phases that began at New Moon).

After Last Quarter Moon, the Moon keeps rising later and later after midnight, and is visible for less and less time before dawn. And its sunlit part that we can see from our viewpoint is a waning crescent, dwindling to a thinner and thinner sliver. More and more of the night is dark for observing stars and other dim objects. Finally, the Moon rises too soon before sunrise to be glimpsed, then it rises and sets right with the Sun—in other words, it is invisible to us and has returned to New Moon. The entire night is again dark for getting our best views of the stars.

Dark Adaptation and Averted Vision

Suppose you are out on a night near New Moon when the sky is very transparent and you are many miles from any significant light pollution. There is hardly any room for improvement in your wonderful sky's condition. But you can still see a lot more in it by using certain visual techniques. The first technique just involves waiting for a natural physiological process to occur properly and knowing how not to interfere with it.

Dark adaptation is a chemical process whereby cells of the retina become more sensitive to faint light after they have been in darkness for a while. You will find that the number of stars you can see after twenty minutes in darkness is much greater than after only five or ten. Most of the improvement takes place in the first twenty minutes or so, but a slight increase continues for many hours. What's really important to know is that even a rather brief and indirect exposure to a fairly bright light can undo some of the dark adaptation. This is why at observational gatherings of serious amateur astronomers, the worst of sins is to point a flashlight in someone's direction. How do amateur astronomers read their maps and manipulate intricate parts of their telescopes or cameras without some artificial light? They use red lights—often special flashlights that have some variety of red filter (for instance, red cellophane) over the light-producing end. What is the point of this? The longer wavelengths of red light do little or nothing to reverse the chemical reaction that makes the *rods* in the retinas more light sensitive. Your dark adaptation is maintained.

Averted vision is a technique astronomers use to try to bring dim light onto the most sensitive parts of the retina by directing the gaze slightly away from the object they want to see. Most of us find ourselves unconsciously falling into doing it—we discover that looking just to the side of a very faint star will make it jump out with greater prominence. But to use this technique as well as possible actually does take some practice. For one thing, different people seem to have slightly different areas of the retina that are most sensitive. Two general things are true for everyone. First, the retinal cells sensitive to color, the *cones*, are found in greatest concentration near the center of the retina and field of vision—which is also the (quite angularly small) area where we are capable of sharp vision. Second, the retinal cells that do not register color but are most sensitive at detecting dim light—the rods—are found at a certain distance away from the center of the retina and field of vision. Thus the trick of using averted vision is to let the sought-after faint light fall on the regions of the retina that have the greatest concentrations of rods.

Binoculars

Binoculars are small, twin, often-inexpensive portable telescopes with wonderfully wide fields of view. Most people should first try to learn to identify the most important stars and constellations, also the planets, all with the naked eye. But then, the next step should probably be to use binoculars, before spending a lot more money on a full-fledged telescope.

Binoculars are designated most basically with two numbers connected by a multiplication sign. Various binoculars are called 7 × 35, 7 × 50, and 10 × 50. The first number is the magnification power—a pair of binoculars can make something look seven or ten (or other) times bigger. The second number is the diameter of each of the main lenses in millimeters (mm). If this is 50 mm, then the lens is about 2 inches wide. The magnification of binoculars is typically much lower than that of astronomical telescopes. With binoculars, you can't see small details on the Moon or the planets, or split close double stars, or see nebulae at a large-enough scale to behold any intricacies of their structure.

The field of view of typical binoculars is usually about 5° or 6°, but some are much less and others are much more. Wide-field models with views of up to 12° or even 13° are available.

Telescopes and Their Three Key Abilities

Almost everyone who has found an interest in astronomy desires to own a telescope someday. Rightfully so, for even though there are great numbers of exciting sky-sights for observers with unaided eyes, telescopes open up immense additional realms of vision and beauty.

This book is not the place to discuss the numerous aspects of telescope selection, use, and care. There are many other books as well as magazine and Internet sources for more detailed information on telescopes. It is worthwhile here, however, to consider some telescope basics, especially as they relate to our central topic, the brightest stars.

First, what are the most important characteristics a telescope should have? That depends very strongly on each person's individual needs. For example, if you are going to have to travel away from city lights frequently in order to view the night sky, the portability of your telescope becomes a major issue. There are, additionally, three most important abilities of a telescope that a user should always consider.

Most vital of all is not, as many people think, the magnification. It is the *light-gathering ability*. Most objects in astronomy are very faint or, at the very least, benefit from being made brighter than they appear to the unaided eye. The light-gathering ability of a telescope depends on the diameter of its primary mirror or lens, which is often called its *aperture*. All other things being equal, the more aperture you can get for your money, the better the telescope. Aperture is also the key to a second crucial ability of telescopes—*resolution*, which is the ability to reveal detail. The wider a primary mirror or lens your telescope has, the more detail it is inherently capable of showing.

The final ability is *magnification*. This can be limited practically by the focal length of the telescope—the distance from the primary piece of optics to where it focuses an image—and the focal ratio—the ratio of the focal length to the aperture. A telescope whose main mirror is 8 inches wide and whose focal length is 48 inches would have a focal ratio or "f number" of 48 divided by 8, which equals 6 (usually written "f/6"). But it is the eyepiece that does the magnifying, and the size of the aperture determines how sharp of an image can form at a given magnification. This is extremely important to know because devious manufacturers of poor telescopes often brag that the small telescope they are selling can provide some enormous magnification such as 500×. And indeed, with the appropriate eyepiece, such a magnification can be achieved. The catch is that unless the telescope also has a large aperture, an image magnified 500× will be a big but hopelessly blurry blob.

The rule of thumb about magnification is that the highest magnification that can give a useful image when "seeing" is excellent is about 50× per inch of aperture. Thus the maximum effective magnification for a 4-inch telescope (a telescope with a 4-inch-wide mirror or lens) would be 4 × 50 = 200× magnification. And you would need a 10-inch mirror or lens (10 multiplied by 50 equals 500×) to be able to get a useful image at 500×—when the atmosphere is exceptionally calm. As a matter of fact, the rule of 50× per inch of aperture really applies only to certain kinds of celestial objects, such as planets and close (that is, tightly paired) double stars. When an experienced observer looks at an object with relatively low surface-brightness—maybe a galaxy or a comet—he or she knows that using high magnification will merely spread the already feeble light of the object over too large an area. Such celestial objects should generally be observed at low magnification (maybe only 10× per inch of aperture) and at no higher a magnification than is needed to make out the object's component features.

Basic Types and Parts of Telescopes

The most fundamental parts of a telescope are its optical system, its tube, and its mount. The available kinds of optical systems divide telescopes into three most important general types: the *reflector*, the *refractor*, and the *catadioptric*.

The reflector uses a *primary mirror* at the bottom end of the tube to reflect incoming celestial light back to a *secondary mirror* that in turn sends the image to an eyepiece tube and eyepiece positioned on the side of the top end of the tube.

The refractor uses a *primary lens* at the top (skyward) end of the tube to focus it eventually down to an eyepiece tube and eyepiece at the bottom end of the telescope.

The catadioptric uses both lenses and mirrors to fold the light path of the celestial object and enable the tube to be shorter than it would otherwise have to be.

Reflectors are known for being relatively inexpensive for their aperture but do need periodic adjustments by the owner to keep their mirrors properly aligned (this process is called *collimation*). Refractors are much more expensive per inch of aperture—especially when the aperture is large, at which point refractors also become much less easily portable than reflectors. But refractors should rarely if ever need collimation and, if they are of high quality, can produce exceptionally sharp images of planets, lunar features, and double stars. Catadioptric telescopes are generally decent all-around telescopes and, because of their folded light-paths, are unusually compact. This, among other things, makes it easier for them to have reasonably light equatorial mounts and take photographs.

Speaking of mounts: in addition to its basic optical system, a telescope must have a mount, eyepieces to add to the optical system, and a finderscope.

A telescope is only as good as its mount. Every Christmas, it's sad to see so many people—and especially children—getting junk telescopes, usually from department stores, for presents. And one of the most common problems with these instruments is a hopelessly shaky mount. When you move the tube to a different target or for other reasons touch a telescope, its mount should be good enough for all vibration to die out within a second.

An *equatorial mount* is one that must have its polar axis aligned to the celestial pole each time you observe. But the benefit is the ease with which you can follow the motion of a celestial object caused by Earth's rotation. With an equatorial mount it is easy to automatically track this motion, which otherwise becomes troublesome—especially at high magnification, where the celestial object will move out of the field of view in a matter of seconds. Being able to keep an object from moving in (and out of) one's field of view is essential for obtaining images of it with long-exposure photography (electronic imaging is more forgiving because it can grab many quick exposures that can then be processed together digitally to produce a single good image).

On the other hand, there are advantages to a mount that permits merely up–down, left–right moving. This is called an *altazimuth mount*. A variety of it with a tube and a rocker box is the deservedly popular kind called a *Dobsonian mount*. The Dobsonian tends to be much cheaper and easier to use than an equatorial mount. It is probably suitable for you if you aren't interested in astrophotography just now (though there are ways to adapt Dobsonians to move equatorially) and if most of your observations will be of stars and other sights beyond our solar system. Even intricate details on the planets or the Moon are not necessarily hard to get with a good, properly collimated Dobsonian telescope.

Eyepieces, also called *oculars*, are small combinations of lenses that are used to magnify the image from the telescope. Every observer should have a selection of eyepieces, making it possible to get different magnifications from the telescope. Veteran observers pay almost as much attention to, and money for, eyepieces as they do for their telescopes. Some eyepieces permit a wider field of view at a given magnification. And obviously, image quality is of utmost importance.

No eyepiece in combination with a fairly large telescope can give a wide enough field of view to make it easy to get celestial objects into the field. Therefore, a *finderscope* with low magnification but very wide field of view needs to be attached to the telescope tube and made parallel to it. After maneuvering a star or other heavenly target to the center of the finderscope's field, it should then be possible to find the target object somewhere in the main telescope's much more narrow field of view.

PART TWO

STARS IN THE UNIVERSE

PARTS, STRUCTURE, DISTANCES, AND MOTIONS IN THE UNIVERSE

5

I n the first part of this book, we were concerned with describing the appearances of things in the heavens, and how they were named and mapped. We mentioned red giant stars and supernovae, star clusters and nebulae, galaxies and double stars, and variable stars of various types. We did not, however, try to discuss much about the true nature of these objects, the realities behind their appearances.

These realities, discovered by science, mostly in modern times, are of great importance even if you're sure you have no inclination to ever become a professional astronomer. How can someone look at a beautiful sight in the night sky and not have some desire to better understand its nature? Of course, not everyone wants to tangle with the mathematics and other challenging intellectual aspects of the science of astronomy. But if we know a little bit more about the realities behind the sights we see, the richness of our experience increases greatly. We may also gain an understanding that will help us anticipate when celestial events we haven't read about are likely to happen or when aspects of them that we otherwise wouldn't know to look for will occur.

The first section of the book introduced us to the brightest stars and then told us how to find and enjoy their appearances in the rich settings of the seasons, the constellations, and the celestial sphere. Now we will learn what stars really are, what causes their different appearances, and how they fit into the great scheme of the universe.

Most of our attention will be to stars themselves. But we start out in this first chapter of the section with a quick introduction to the structure and workings of this universe in which stars play such a fundamental and integral role.

The Sun and Its Solar System

A *solar system* is a collection of a *sun*—that is, a star—and all the objects that are not essentially self-luminous that revolve around it, directly or indirectly. Several different kinds of such objects may orbit a sun.

A *planet* is a major body that orbits a star (such as our Sun) directly and that is not self-luminous. (But note the ambiguity of the word *major* in my sentence. In recent years, the definition of planethood has become controversial due to the nature of Pluto and the finding of objects that resemble it, as well as other discoveries. For instance, astronomers have detected planet-size non-self-luminous worlds beyond our solar system that travel through space without the company of any star.)

Most of the planets have bodies orbiting them that are therefore called *moons* or *satellites* (natural satellites, as opposed to the artificial satellites that humankind builds and launches into orbit around Earth). Earth's one large natural satellite is called the Moon.

Several kinds of objects much less massive than planets revolve directly around the Sun. Those that are mainly rocky permanent bodies are called *asteroids* or *minor planets*, and they mostly revolve around the Sun between the orbits of the planets Mars and Jupiter. Some do cross the orbits of Earth, however, posing a threat for rare but devastating collisions with our planet.

Comets are mainly icy and more temporary bodies. They often get into elongated orbits that bring them near enough the Sun to start vaporizing (actually *sublimating*—going directly from solid to gas in the low-pressure near-vacuum of space).

Kuiper Belt Objects (KBOs) (some of which are also called *Trans-Neptunian Objects* or TNOs) are probably mixtures of rock and ice (perhaps often predominantly ice?) orbiting the Sun around and beyond the orbit of Neptune (these objects resemble Pluto but in 2006 the International Astronomical Union ruled that Pluto and the other KBOs should not be considered planets).

Meteoroids are iron or rocky (or at least silicate) objects found throughout the solar system ranging in size from microscopic specks up to the dimensions of the smallest asteroids. The diameter that marks the dividing line between the largest meteoroids and the smallest asteroids is arbitrary and not officially agreed-upon (any of these rocky bodies more than a few hundred meters or yards wide might likely be called a small asteroid). When meteoroids enter Earth's atmosphere, they burn up and produce streaks of light we call *meteors*. Very rarely, they each the ground and are then called *meteorites*.

Scientists are identifying more and more solar system objects that seem to be transitional forms between, or otherwise hybrid forms of, those listed here.

Stars, Nebulae, and Star Clusters

When we make the leap from our solar system out to the stars, the increase in distance is truly staggering. We'll get to that matter—and the question of how astronomers have been able to determine these distances—a bit later in

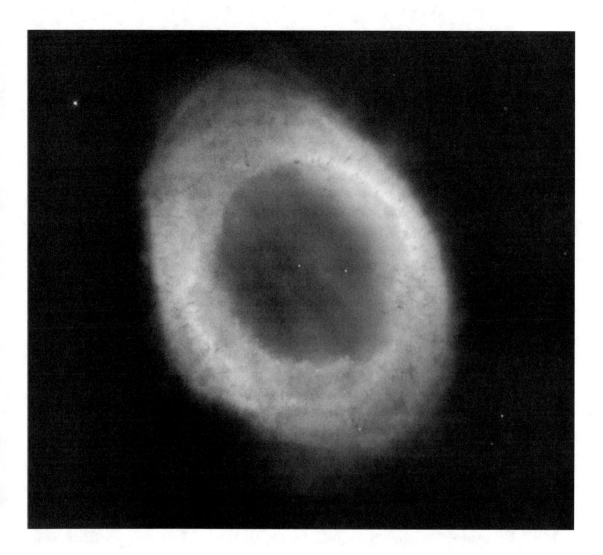

this chapter. First, we will continue with our inventory of the objects found as we widen our journey beyond the solar system to the stars, galaxies, and the universe.

A *star* is a self-luminous ball of gas that generates energy through nuclear fusion (though in the case of white dwarf stars, the fusion occurred in the past and the star merely shines on for a vastly long time). We'll discuss the different kinds of nuclear fusion that power stars later in this book. We'll be looking at the lives and deaths of stars and find that the final stage in the lives of very massive stars is to implode and explode as a *supernova* and then collapse into the famous bizarre objects known as *black holes.* We'll also examine the many different varieties of stars—including the different types of double

The Ring Nebula (M57), the most famous planetary nebula, as seen by the Hubble Space Telescope.

stars and variable stars. But in our current tour of the universe and the motions within it, our next destination is the clouds in which stars are born and that they, in death, produce—nebulae. **Nebulae** are clouds of gas and dust between and around the stars.

Diffuse nebulae, such as the Great Orion Nebula (M42), often large and irregularly shaped, are clouds in which new stars are being born. There are two major kinds of diffuse nebula. An *emission nebula* is a diffuse nebula in which the gas itself glows (mostly pink from hydrogen) from being energized by the radiation of nearby hot stars. A *reflection nebula* is a diffuse nebula in which the gas glows (usually blue) by merely reflecting the light of nearby stars. Many diffuse nebulae—such as M42, M8 (the Lagoon Nebula), and M20 (the Trifid Nebula)—have some areas that are emission regions and others that are reflection regions. The colors of these nebulae usually take at least a medium-size telescope to begin to glimpse and the hues are never anywhere near so vivid visually in the sky as they are in long-exposure photographs.

Dark nebulae are clouds of gas and dust that have no stars close enough to them to make them emit or reflect light. Therefore, we see dark nebulae only where they are foreground silhouettes against luminous nebulae (as in the case of the famous Horsehead Nebula) or where they are apparent as a sudden extreme reduction in the number of stars in a particularly rich star field.

Planetary nebulae are the clouds of gas and dust released by dying stars. Compared to diffuse nebulae, they are small. But their light is usually intense, and even their color can be vivid because they are energized by intense short-wavelength radiation of the *central star* that gave rise to them. The most famous examples—the Ring Nebula (M57) and the Dumbbell Nebula (M27)—are, like all their brethren, visible as somewhat geometrical patterns, the products of sometimes multiple outbursts from particular regions of a rapidly rotating star. Let me stress that planetary nebulae have nothing in particular to do with the formation of planets. They gain their name from the fact that many of them, when viewed in small or medium-size amateur telescopes, somewhat resemble the blue-green fuzzy disks of the planets Uranus and Neptune.

Supernova remnants (SNRs) are clouds, such as those of the Crab Nebula and the Veil Nebula, that are left over from the massive star explosions known as **supernovae**.

A **star cluster** is a true grouping of stars in space (not just an apparent one in the sky) in which the component members are more numerous and generally much farther separated from one another than in a double-star system (or multiple-star system—Castor is sextuple). The two major kinds of star clusters are *open clusters* (sometimes called *galactic clusters*) and *globular clusters*.

Before we describe what these two types are like, we should note two kinds of groupings that somewhat resemble clusters.

Moving groups and *associations* are more loosely bound than clusters. Moving groups may in some cases be previous star clusters whose members have finally drifted far enough apart to be on the verge of losing any locational connection with one another. The latter (associations) are groupings of young stars that happened to form in the same active star-birth area of a spiral arm of a galaxy. The fate of associations will be to break up in a matter of only a few million years rather than in hundreds of millions or a few billion years, like open star clusters.

Open clusters are irregularly shaped, relatively loosely packed groupings of typically a few dozen to as many as a few thousand stars. They owe their name to the fact that there is more open space between the members than is the case with the other kind of star cluster, globular clusters.

Globular clusters are roughly spherical, densely packed groupings of typically a few hundred thousand to a few million stars.

Open clusters and globular clusters are fundamentally different entities. There are thought to be a few tens of thousand open clusters in our galaxy, but probably no more than about two hundred globulars. Open clusters are located almost entirely in the equatorial disk of our galaxy (where the Sun itself is located). Globular clusters form a roughly spherical halo around the galaxy, some of them at great distances from the galactic center and far above or below the galaxy's equatorial disk. Open clusters are made up of *Population I stars*, younger stars of later generations of star formation (formed in part from the gas left over by supernova explosions of earlier-generation stars). Globular star clusters consist of *Population II stars*, much older, first-generation stars, some of which are probably almost as old as our galaxy. The central hub of our galaxy probably consists largely of Population II stars and seems almost like a mega-globular-cluster itself.

Galaxies and the Universe

A **galaxy** is a vast congregation of often billions or even hundreds of billions of stars. There are three fundamental types. *Spiral galaxies*, like our own Milky Way galaxy, are large pinwheels of stars consisting of a roughly spherical central hub and an equatorial disk of stars, gas, and dust that includes spiral arms. *Elliptical galaxies* seem almost like the central hubs of spiral galaxies without the disk or arms (they appear to consist of mainly Population II stars)—though some ellipticals are even bigger than spirals. *Irregular galaxies* are asymmetrical in shape and are generally small galaxies with fewer stars

The spiral galaxy M51 as seen by the Hubble Space Telescope.

than spirals and ellipticals (they can still have plenty of Population I stars, star clouds, and nebulae—almost as if they were detached sections of spiral arms).

Galaxy groups and *galaxy clusters* are collections of galaxies. The *Local Group* of galaxies consists of the Milky Way, M31 (the Great Andromeda Galaxy), M33, and several dozen lesser galaxies, many of them companion or satellite galaxies of the Milky Way and M31. A galaxy cluster such as the Virgo Galaxy Cluster is tremendously bigger, including thousands of galaxies.

Across the inconceivably vast distances of intergalactic space, two even more energetic kinds of objects than supernovae should be mentioned. The first is *gamma-ray bursters*, some of which are the most energetic events in the universe for the brief time they occur—and some of which may be caused by a collision of black holes. The second is the most enduringly energetic objects in the universe. These are **quasars** (originally called *quasi-stellar objects* because they appear as starlike points of light). Quasars may be only a few times larger than our solar system but can nevertheless emit more light and other energy than many entire, vast galaxies do. None of the quasars seems to exist in our era of the universe, but they are so bright that they can be seen from billions of light-years away—in other words, the light we see reaching us from most of them left them billions of years ago, in most cases before the Sun and Earth even formed. Quasars are still mysterious but the leading theory is that they are a peculiar kind of galactic core powered by massive black holes.

The *universe* is all that physically exists—or at least all that exists and that the laws of nature allow us to get information about, despite the universe's expansion. Let's back up—way back, to a time 12 to 14 billion years ago. At that time, most cosmologists believe, time and space began with the event called *the Big Bang.* It was not really an explosion but an expansion—and not of objects and energy into space but of space itself with matter and energy already in it. The expansion continues to our present time. It is demonstrated by the fact that all galaxies (except for those gravitationally bound with our Milky Way as the Local Group) are moving away from us. The farther a galaxy or other object is from us, the greater the sum of its and our velocity in the expansion of the universe—and thus the faster it is receding from us. When the velocity of recession at a certain distance—something like 15 billion light-years—reaches the speed of light, even light and other electromagnetic radiation—the fastest thing in the universe—cannot travel fast enough for it to ever reach us. We are therefore, it seems, fundamentally incapable of receiving any direct information about any matter or energy beyond that limit.

Interplanetary Versus Interstellar Distances

Billions of light-years? Such distances we cannot truly grasp. But in our tour outward in the structure of the universe I have deliberately left out most of the figures about distances. I want us now to go back and try to comprehend first the distances within our own solar system, then the gulfs between the main figures of this book—the stars.

The distances between the planets utterly dwarf those we are used to on our own world. The Moon orbits an average of about 30 times Earth's diameter away from us. That's a little less than 10 times as far as a trip all the way around our world. But the closest that Venus, our closest planetary neighbor, comes is a little more than 100 times as far as the Moon. The closest that Pluto, famed for its remoteness, ever comes to Earth is about 100 times farther than the close approach of Venus. Thus, although our astronauts managed to visit the Moon at the end of a journey that was just a few days long, Pluto never comes closer than 10,000 times as far away.

But these distances are as nothing compared to those between the stars.

Perhaps the best way to convey first the hugeness of the solar system and then the ultra-hugeness of interstellar space is with astronomy writer Guy Ottewell's famous demonstration, the "1,000-Yard Model" of the solar system.

In this activity, teachers and students use a soccer ball or ball of similar size—about 8 inches wide—to represent the Sun. How big is Earth? The size of a peppercorn—a minute object only about .08 inches wide. And the distance one has to walk from the soccer ball to reach the peppercorn-size Earth in its orbit? Twenty-six yards! That sounds impressive, but you have to actually get objects this size and do the walk yourself to see how amazing it really is. Even the most knowledgeable astronomer, who thought he knew how vast the difference was by mental calculations, is surprised when it is there before his or her very eyes.

But the real shockers are yet to come. Mars is another 14 yards farther from the Sun than Earth is. But when the teacher/presenter starts pacing off the distance between Mars and Jupiter, is when the gasps start coming. We've already walked 40 yards to get from the Sun to Mars. Now the walk from Mars to Jupiter is another 95 yards—almost the length of an (American) football field. But Jupiter is the fifth of the traditional nine planets, so we're halfway out to the edge of the solar system, right? No, the gaps between the orbits of the outermost planets are much larger. The halfway point comes at about the orbit of Uranus, with only two traditional planets left. The average distance from a soccer-ball-size Sun to a pinhead Pluto (one-hundredth of an inch wide) is 1,010 yards.

Kuiper Belt Objects and vast numbers of comets are much farther out than Pluto yet still under the gravitational sway of the Sun. But all save a few of these bodies would have to be represented by objects too small to see with the naked eye. So in the 1,000-yard model, how much farther than the orbit of Pluto would you have to go to reach the nearest star system (the Alpha Centauri system)? 5 miles? 10 miles? Maybe 100 miles?

If the soccer ball (our Sun), the peppercorn, and the pinheads (Earth and its fellow planets) were scattered in a large field in New York, we would have to picture all of North America completely deserted—except for a similar field with three balls and perhaps a few more bits of nuts and rubble—somewhere in Alaska. That would be the Alpha Centauri star system, the nearest to our own.

Interestingly, of all the transitions—from on-Earth distances to the Moon, the Moon to the planets, the planets to the stars, the stars to the galaxy, and the galaxy to the universe—by far the greatest ratio is that of stellar distances to planetary. The Milky Way's satellite galaxies are little more than one Milky Way length away from the Milky Way. M31 is only about 25 Milky Way lengths away. The center of the Virgo Galaxy Cluster is only about 600 Milky Way lengths away. Even the farthest galaxies and quasars ever observed are only about 12,000 or 13,000 Milky Way lengths away. By comparison, Earth's distance from Alpha Centauri is about 270,000 times farther than Earth's distance from the Sun.

Units of Distance in Astronomy

The distances in astronomy are so vast that they require special terminology. Even within the solar system, the distances from the Sun to the outer planets is measured in a few *billions* of miles or kilometers, unwieldy figures. More manageable and useful for many purposes is the *astronomical unit*, or *AU*, a unit of distance that is the average distance from the Sun to Earth—about 150 million kilometers, or about 93 million miles. The average distance of Neptune from the Sun is about 30 AU; of Pluto, about 39 AU.

But even the AU is inadequate for the distance to the stars. Remember, the nearest star system to our own is more than a quarter of a million astronomical units away.

For an appropriate unit to measure interstellar distances, astronomers must turn to the fastest thing in the universe, light (or any form of electromagnetic radiation). Light in a vacuum travels at 186,300 miles per second or about 300 kilometers per second. The unit of distance, not time, called a **light-year** is the distance that light can travel in a year. Light only takes about 1¼ seconds to travel from the Moon to the Earth and just over 8 minutes to fly from the Sun to the Earth. When we see Neptune, we are actually seeing light that left it about 4 hours ago (it is 4 *light-hours* distant from Earth). 1 light-year equals about 63,000 AU. So the nearest star system is 4.3 light-years away. The most remote of the 1st-magnitude stars may be as many as 2,600 light-years away.

You will also sometimes read in astronomy texts about a unit of distance a little bit larger than a light-year. The **parsec** is the distance at which an object would have an annual parallax of one arc-second. **Parallax** is the change in apparent position of an object in relation to a more distant background when the object is seen from two different viewing points. The closer the foreground object, the larger the parallax. You can check this out for yourself by holding your finger at arm's length in front of you and looking at it first with one eye (close the other) and then with the second eye (closing the first eye). The change in the viewing point of your right eye and left eye will make your finger shift position against the more distant background of a wall. But if you put your finger closer to your eye and try this experiment, you will notice a much larger shift in position of your finger—a much larger parallax. Stars are so remote that even using the two opposite sides of Earth's orbit as your different viewing points causes only a tiny change in position of even the nearest stars against the background of much more distant ones. No star system is close enough for us to observe a position shift of as much as one arc-second when observed six months apart (that is, from opposite sides of Earth's yearlong orbit). Even Alpha Centauri is about 1.3 parsecs from Earth. One parsec equals 3.2616 light-years.

By the way, until the very end of the twentieth century, it was difficult to get precise enough measurements of stars' positions to measure with much precision the parallax of stars even as close as 50 or 100 light-years away. But in the 1990s, the Hipparcos satellite, working above the unsteady atmosphere of our planet, was able to make much more precise measurements of positions and parallaxes. Before the Hipparcos satellite, distances even to stars as relatively close as 80 light-years were known only to an uncertainty of about 25 percent; after Hipparcos, the uncertainty of the distance to stars 80 light-years away dropped to only 2 percent.

Even Hipparcos is not capable of determining fairly accurate parallaxes for stars more than a few hundred light-years away. Astronomers have to rely on other methods to estimate how far away more remote stars are. A key is, through spectroscopic or other study, figuring out the nature and therefore the true brightness of a star. For, once we know the true brightness, we have only to compare it to the *apparent* brightness of the star to figure out how far from us it must be. If a certain kind of star always has the same true brightness, then all examples of that sort of star can have their distances correctly calculated. The most outstanding case is that of Cepheid variable stars. In these stars, it was found, the period of brightness variation was proportional to the true brightness—the longer the period, the greater the true brightness. Spectroscopic study of double-star systems can also establish the mass and spectral type of the component stars and then use the "mass-luminosity relation" to calculate a distance—but these are matters for our next two chapters.

Luminosity and Absolute Magnitude

What I want to establish here now is the units in which the true brightness of stars is given. The first of these is **luminosity**. In the rest of this book, when we speak of the luminosity of a star, we are not using a pleasant synonym for "brightness." We are talking specifically about the true brightness of a star as a sun in space. Luminosity is often given in terms of our own Sun's true brightness. For instance, in the tables at the back of this book, Sirius is listed as having a luminosity of 26—meaning it is 26 times as luminous as the Sun.

A second useful measure of the true brightness of stars is absolute magnitude. We became acquainted with the magnitude system in chapter 1. But so far we have been talking only about the **apparent magnitude** of stars—the brightness they appear to have as seen in our sky. That brightness depends both on the star's true brightness and its distance from us. **Absolute magnitude**, like luminosity, is true brightness, for it is the apparent magnitude a star would have if placed at a standard distance of 10 parsecs (which is about 32.616 light-years). At such a distance, the two stars of greatest apparent brightness would be far less impressive. Night's brightest star, Sirius, has an apparent magnitude of –1.44 but it is only 8.6 light-years away, so its absolute magnitude is +1.5. Day's brightest star, the Sun, has an apparent magnitude of –26.75 but it is only 8.3 light-minutes away, so the absolute magnitude of the Sun is +4.8. If our Sun was about 32.6 light-years away, it would be a rather faint naked-eye star, outshined by about a thousand others in our sky.

On the other hand, most stars we see with our naked eye are, unlike the Sun and Sirius, farther than 32.6 light-years from Earth. Their absolute magnitudes are therefore brighter than their apparent magnitudes. An extreme example is Deneb. Deneb's apparent magnitude is +1.25. But even if it is "only" 1,500 light-years away (many estimates place it much farther, perhaps as distant as 2,600 light-years), its absolute magnitude is –7.5. That means that if Deneb were 32.6 light-years from us, it would shine at –7.5—similar in brightness to a crescent Moon.

Motions of Stars in Space

We've listed the major parts in the structure of the universe and discussed the distances between those parts. All that remains for this chapter is to explain how the parts move. The fact that objects in the solar system move in more or less (in some cases, very much less) circular orbits around the Sun is well known. The movements of galaxies within our Local Group or of even larger-scale motions—other than the expansion of the universe—are mostly poorly known. So it's only the motions of stars relative to one another that needs some explanation here.

The ancients could see that the constellations of each season came and went. Hipparchos even discovered the precession of the equinoxes. But everyone knew that the *patterns* of the stars—that is, their positions relative to one another—didn't change. At least, not noticeably in the course of a human lifetime or even hundreds of years.

But in 1718, Edmund Halley published his finding that the positions of three stars—Arcturus, Sirius, and Aldebaran—were slightly different from what they had been in the ancient star catalogs. He had discovered the *proper motion* of stars.

Each star has two observable components of its true motion (ones that are not merely effects of our observing platform Earth's rotation, orbital motion, precession, and other movements). One component is **proper motion**—a star's movement in some azimuthal direction (north or southeast or other) in relation to the background of much more distant objects. If we also know the distance to the star, then the amount of its proper motion can give us the *tangential velocity*—the transverse motion in space in kilometers or miles per second.

The other observable component of a star's true motion is its *radial velocity*—the part of its movement that is either approaching or receding from us. That can often be learned by noting if lines in the spectrum of the star's light are either *redshifted* or *blueshifted*—transplaced toward either the red or the blue end of the spectrum. This shifting is an example of the *Doppler effect*. Just as the sound of an approaching car's horn is shifted to a lower pitch, an approaching star's light is shifted toward a lower (shorter) wavelength—blue. And just as the sound of a receding car's horn is shifted toward a higher pitch, a receding star's light is shifted toward a higher (longer) wavelength—red. (It was by their redshifts that it was discovered that the galaxies are receding from us, that the universe is therefore expanding—and the amount of the redshift determines the speed of recession, which in turn enables us to calculate the distances of galaxies.)

If we know both the transverse and radial motions of a star, we can calculate the true motion in three dimensions, the true or *space velocity* of the star.

The space velocities of most stars turn out to be lower than we might expect. But this is because most stars we see are part of the Population I kind in the equatorial disk of our galaxy and, as such, are pursuing speeds and directions around the galactic center that are not greatly unlike those movements of our own Sun. In other words, most stars we see are, like the Sun, part of the mainstream traffic circling the galactic center. Only a few stars—most notably Arcturus—are diving or climbing past the rest of us in the equatorial disk of the galaxy. They, too, circle the galaxy's center but in orbits considerably inclined to the equatorial plane. Therefore, these stars have larger space velocities relative to our Sun (and to most stars in the solar neighborhood) as they swoosh past us pursuing their very different routes.

The point on the celestial sphere toward which the Sun and its solar system are headed is called the *Apex of the Sun's Way*. The point from which the Sun and the solar system are receding is called the *Antapex of the Sun's Way*—or, more eloquently, the *Sun's Quit*. We will discuss these points in the sky when we profile the two very bright stars nearest to them—Vega (near the Apex) and Sirius (near the Antapex).

THE VARIETIES OF THE STARS

The variety of stars is astonishingly great. Stars come in a wide range of colors and temperatures. They can be as large as red giants, hundreds of times larger than our Sun. They can be as small as white dwarfs less wide than the Earth, neutron stars about the size of a city, and black holes (if we can still call them stars), which can be considered infinitely small. There are also double and multiple star systems in a great number of positional arrangements and many combinations. There are dozens of different varieties of variable stars with periods of brightness variation ranging from many years to small fractions of a second. There are stars that in the place of our Sun would not greatly outshine our Moon. There are other stars so luminous that in one night they release more light than our Sun does in a century.

As diverse as stars are, we will find that most of the kinds of stars and stellar attributes that exist are represented by at least one of the 1st-magnitude stars or its companions (or will be in the eventual evolution and death of the star).

Before we can try to understand what *causes* stars to be so different and piece together the stages of a star's life and death, however, we must examine the many different kinds of stars that we can observe or deduce from other data. Let's begin the adventure.

Double and Multiple Stars

A **double star** is a star that, upon closer or more sophisticated examination, proves to be two stars. The term can be loosely extended to include systems of three, four, or more stars, though a system with three or more stars is often called a **multiple star**. Astronomers believe that most stars are members of either double- or multiple-star systems. Our Sun seems to be single, but there is still a chance that it has a distant and very dim companion that we will someday discover by virtue of its having a huge proper motion in the heavens.

Not all apparent pairings of stars prove to be a real pairing in space. **Optical doubles** are apparent pairs of stars in which the two stars are really objects at greatly different distances from us that just happen to lie along

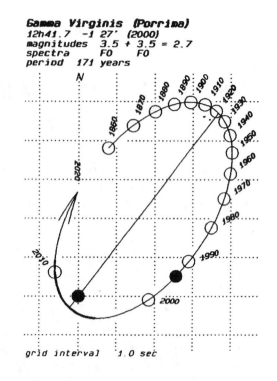

Gamma Virginis (Porrima)
12h41.7 -1 27' (2000)
magnitudes 3.5 + 3.5 = 2.7
spectra   F0   F0
period   171 years

grid interval 1.0 sec

A diagram of a double star orbit. The companion star's position for 1994 is indicated by a filled-in circle. Note that this double is now opening from tightest separation.

nearly the same line of sight. Alpha Capricorni (for naked eye and binoculars) and Delta Herculis (for telescopes) are interesting examples.

The brighter, more massive star in a double system is called the *primary*. The less bright, less massive member of the system is the *secondary* or *companion*. Most double stars are **binary stars**—star systems in which the component stars actually orbit around each other (or, more strictly speaking, around a common center of gravity called the *barycenter*). But some companions are at great distances—even up to a light-year or more—from the primary and are called *common motion doubles* because they only share a common proper motion through space.

Some double stars are *visual doubles*, pairs of stars that can be resolved into two separate points of light by a telescope. *Spectroscopic binaries* are doubles too close together to split in telescopes but whose *duplicity* (doubleness) can be detected by its effect on the spectrum of the double star's light. A pair that is approaching us in space may show a redshift in its spectral lines due to the fact that one of the component stars is also moving away from us in the course of its circling around the other component.

Observers need to know the separation of a double star and the position angle of the companion. The position angle, or P.A. is the azimuth, with 0° as north, 90° as east and so on. In most binaries, the position angle changes at least slightly in the course of years or decades. But some stars have such vast orbits around each other that their separation and position angle stay relatively fixed over a few hundred years—leading to their being called *relfixes*.

Variable Stars, Novae, and Supernovae

Variable stars are stars that vary in brightness. Observers need to know the *period* over which the changes in brightness occur and the *amplitude* (amount) of the changes. They also need maps showing *comparison stars* of fixed brightness with which the variable star's brightness can be compared and estimated during the course of its changes.

There are dozens of kinds of variable stars. We need to discuss only a few types here. The first and most fundamental division is between two overarching kinds: extrinsic variables and intrinsic variables.

Extrinsic variables are those that change brightness due to an external cause—an eclipse of a primary star by its companion. Such stars are **eclipsing binaries**, whose brightest and most famous example is the 2nd-magnitude Algol in Perseus. They are such close doubles that we cannot split them visually—but the point of light's dips in brightness reveal the system's duplicity.

Intrinsic variables are those that change brightness due to internal causes. These can be broken down into pulsating variables and eruptive or explosive variables.

First, let's consider *pulsating variables*, stars that change their luminosity due to repeated changes in size, temperature, and color. *Irregular variables* (usually red giant stars) change brightness with no predictable pattern. *Semiregular variables* are red giant stars that have some periodicity but also unpredictable brightness fluctuations superimposed on the periods. Betelgeuse is probably an example.

Long-period variables are red giants with amplitudes (brightness ranges) greater than those of semiregulars and with periods that for some stars is as short as about two months, while for others as long as about two years. The time of maximum may be off prediction by up to a few weeks, and the maximum itself may occasionally be much brighter than usual and last for several weeks. Long-period variables are the most common kind of variable star, with about a quarter of all variable stars falling into this class. The brightest and most famous long-period variable by far is Mira (Omicron Ceti)—in fact, long-period variables are sometimes called Mira types or even just "Miras." The second most interesting long-period variable is probably Chi Cygni, which usually ranges from about 12th up to 5th magnitude in a period of slightly over 400 days. Chi Cygni reached its brightest peak in more

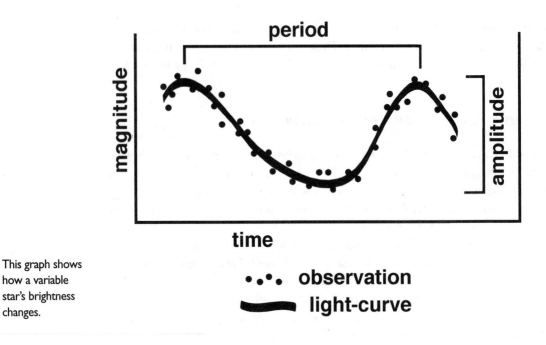

period

magnitude

amplitude

time

•.•. observation
━━━ light-curve

This graph shows
how a variable
star's brightness
changes.

than 140 years—almost magnitude 3½—in the summer of 2006. It is located only 2½° from Eta Cygni, the magnitude 4.0 star in the Swan's neck, and when unusually bright seems to make a kink in that neck.

Cepheids (or at any rate Classical Cepheids) are yellow supergiant stars that have very precise amplitudes and periods. The latter range from about 1 to 70 days. As mentioned in the previous chapter, the length of a Cepheid's period is usually proportional to its luminosity (the longer the period, the greater the luminosity). This *period-luminosity law* enabled astronomers to calculate the distances to neighboring galaxies with the help of Cepheids that were in them. Thus the Cepheids, named for Delta Cephei, are sometimes called "flickering yardsticks," "measuring-sticks of the universe," and "standard candles." Delta Cephei dims from magnitude 3.5 to 4.4 in about four days, then back to maximum in about a day and a half. The length of the entire period is 5 days, 8 hours, and 48 minutes.

Let's now consider a few of the types of *eruptive* or *explosive variables*, stars that have more violent causes for their variations in brightness.

Flare stars (also called *UV Ceti* stars) are faint red-dwarf stars that occasionally have brightenings of up to several magnitudes in a few minutes due to flares, magnetic disruptions that spew atomic particles and energy into space.

Recurrent novae are stars such as T Coronae Borealis ("the Blaze Star"), which brighten by a lesser amount than true novae but still do so by about 6 or 7 magnitudes in a few days, fading back to near their original brightness in a matter of weeks. Unlike true novae, these stars undergo repeated brightenings, as often as every few decades or each century. (Actually, it's possible that true novae also have repeat performances—but, if so, they probably happen after lapses of thousands of years, for no such repeats have yet been recorded in human history.)

Novae (singular *nova*, Latin for "new") are stars that brighten rapidly by 7 to 16 magnitudes and then dim back toward their original brightness, typically in a matter of weeks. Novae occur when a white dwarf steals hydrogen gas from a close binary companion, and this material builds up in an *accretion disk* that eventually falls on the extremely hot surface of the white dwarf, leading to a limited explosion. The excess material is then tossed off in shells of gas.

Supernovae are (with one kind of exception) giant, massive stars that collapse and explode, in the process brightening by 15 to 20 or more magnitudes, possibly leaving a glowing cloud called an SNR—a supernova remnant—and definitely leaving a core that has become a neutron star or a **black hole**. (The exception to this basic scheme is a kind of supernova that results from a more extreme version of the nova scenario in which the infalling material from a binary companion raises the white dwarf's mass to greater than 1.4 times that of the Sun.) We'll look much more closely at supernovae, neutron stars, and black holes in the next chapter.

The Spectrum of Starlight

In the nineteenth century, one of the greatest breakthroughs in the history of astronomy came with the invention and use of the *spectroscope*. A spectroscope utilizes a slit and a prism to spread out the light of a star into its full spectrum of different wavelengths. Those wavelengths register on the eye as the colors red, orange, yellow, green, blue, indigo, and violet in order of decreasing wavelength.

By the way, the spectrum of electromagnetic radiation, which includes visible light, continues beyond violet with wavelengths shorter than those of visible light—ultraviolet, X-rays, and gamma rays—and beyond red with wavelengths longer than those of visible light—infrared, microwaves, and radio waves. Some astronomical bodies produce radiation in many of these wavelengths. In the past few decades, those wavelengths have become increasingly observable by improved detectors and by satellites being able to make the observations above Earth's atmosphere, which blocks some of the wavelengths from reaching us at the surface.

But let's return to the spectroscopic revolution of the nineteenth century and to the spectrum of visible starlight.

The spectrum contains detailed information about the chemical composition of a star, which provides tremendous insight about the star's nature. The *continuous spectrum* of a star is the colors, the different wavelengths, produced by gas under the high pressure and temperature inside stars. The *emission spectrum* is bright lines produced at certain wavelengths by elements and compounds in a glowing gas under low pressure (in the outer parts of a star). The *absorption spectrum* is dark lines at certain wavelengths where other substances in tenuous gas (typically, the "atmosphere" of the star) absorb light. Comparing the patterns of bright and dark lines in a star's spectrum to those of a spectrum of various chemicals in the laboratory enables us to determine which chemicals are present in the star. In this way it becomes possible to learn much about stars and to group them into spectral types.

Spectral Types

Spectral types correspond closely to the surface temperature and color of stars. The hotter a star's surface, the shorter (and more energetic) the average of the wavelengths of the light it emits. Thus, the coolest stars shine red, slightly hotter stars are orange, still hotter are yellow, even hotter are white, and hottest of all are blue-white. (The actual hues of stars are not quite so pure or simple as these colors suggest, but the latter will serve for the purposes of discussion and for initial observations).

The basic Harvard system of spectral classification is credited as having been formulated by E. C. Pickering in 1890 but was later developed by his female observatory assistants, W. P. Fleming, Antonia Maury, and especially Annie Jump Cannon. The earlier alphabetic-order use of letters for the spectral classes had to be rearranged into the following order, working from hottest to coolest stars: O, B, A, F, G, K, M. The old mnemonic for remembering this is the somewhat sexist "Oh, Be A Fine Girl, Kiss Me." Presumably G could be made to stand for Guy by a female user of the mnemonic. Later, classes R, N, and S were added at the red end and could be appended to the mnemonic as ". . . Right Now, Smack!" But R, N, and S partly overlapped with M and, in the later part of the twentieth century, were combined into C, carbon stars. W has been used for the peculiar Wolf-Rayet stars which are certainly near the hot end and so could be appended to the front of the mnemonic as "Wow!" Today, we often see subtypes WC and WN for the Wolf-Rayet stars. Odd additional classes, such as Q for novae and P for gaseous nebulae (which aren't even stars), are also sometimes seen. But our focus here will be on the most important, universally accepted classes: O, B, A, F, G, K, M.

Type O stars shine at temperatures of around 35,000 K. (K is for Kelvin—essentially the Celsius system of temperature but started from absolute zero, where all molecular motion ceases; thus, 0 K (no degree symbol used) = –273.16° C). Type O stars are blue-white stars that are very massive and luminous and live shorter lives than any other kind of star. They are also the rarest of all the seven major spectral types. Their spectra contain lines from ionized helium, nitrogen, and oxygen. Examples are few but include several stars in Orion (Iota Orionis is one).

Type B stars shine at around 20,000 K and, like O stars, are blue-white, massive, and luminous. They display strong helium lines. There are so many in Orion that B stars are sometimes called "Orion stars." Examples include Rigel, Regulus, and Spica.

Type A stars have surface temperatures of around 10,000 K. They tend to be much more luminous than the Sun but much less luminous than type O or B stars. They have no helium lines but have strong hydrogen lines in their spectrum. They are often called "Sirian stars" because Sirius is an example (so is Vega).

Type F stars burn at about 7,000 K and appear yellow-white. They display weak hydrogen lines but show strong calcium lines. Examples include Procyon and Canopus.

Type G stars have surface temperatures of about 6,000 K. The hydrogen lines are weaker but there are stronger lines of many metals. Examples include Capella and the Sun, leading to these being called "solar-type stars."

Type K stars have temperatures of about 4,000 to 4,700 K. They look slightly orange. Their spectra show faint hydrogen lines, strong metal lines, and hydrocarbon bands. Examples include Arcturus and Aldebaran.

Type M stars are supposedly red but typically appear a fairly deep orange. Their surface temperatures are about 2,500 to 3,000 K. They have many strong metallic lines and wide titanium oxide bands in their spectrum. Prime examples are Betelgeuse and Antares.

Each spectral type has within it smaller subtypes numbered from 0 to 9 in order of decreasing temperature. In other words, an A9 star is followed by—is a bit hotter than—an F0 star.

Luminosity Classes

The current version of the Harvard system most often used is called the MKK (Morgan-Keenan-Kellman) system. In this system, the letter and the Arabic numeral of the spectral type and subtype (such as G2) has added to it a Roman numeral for the *luminosity class* to which a star belongs. This is derived not by just calculations of a star's luminosity based on its known distance and apparent brightness, but on a particular feature of a star's spectrum. Bright stars tend to be larger and have more tenuous outer regions, and this leads to *narrower* spectral lines (pressure broadens spectral lines). Thus, on the basis of the spectral lines, stars are categorized in one of seven luminosity classes.

The luminosity classes range from I (the most luminous) to VII (the least luminous). Each class has a title, too. Class I is *supergiants*, class II is *bright giants*, class III is *giants*, class IV is *subgiants*, class V is *dwarfs*, class VI is *subdwarfs*, and class VII is *white dwarfs*. (In practice, though, stars are much less often assigned to classes II and IV than to the other classes.)

Calling luminosity classes by names that seem to focus on size can be initially confusing for beginners. This is especially so because it can distract from a more basic nomenclature for different types of stars that is more useful for discussing the fundamentals of how a star evolves throughout its lifetime. This nomenclature speaks of red giants, blue giants, red dwarfs, and white dwarfs. When I've written earlier in this book that Betelgeuse is a red giant and Rigel a blue giant, I've been referencing the earlier, simpler nomenclature. Betelgeuse and Rigel are both so luminous that they are class I stars, and as such are now technically classified as "supergiants." But it is less important when discussing the stages of most stars' lives to distinguish whether a star has become a supergiant, a bright giant, or a giant, than to establish that it has reached the "red giant" stage.

By the way, within class I, a distinction between very bright and even brighter is made by placing the former (less bright) in class Ib and the latter (more bright) in class Ia. Rigel is placed in class Ia and Canopus Ib—with Betelgeuse intermediate in luminosity at Iab!

So the full spectral label, including luminosity class, for Rigel would be B8 Ia. Betelgeuse is M2 Iab. And Canopus is A9 Ib.

In the MKK system, additional lowercase letters may be added to the luminosity class to note special features in some stars' spectrum. So an *e* means "emission lines;" an *m*, "metallic lines;" an *n*, "nebulous lines;" a *p*, "peculiar spectrum;" a *q*, "lines with blueshifted absorption and redshifted emission, indicating the presence of an expanding shell;" and a *v*, "variable spectrum."

What is the Sun's spectral label in the MKK system? It is a G2 V star. This makes our Sun a yellow dwarf. But there is another name for class V stars that sounds better: "the main sequence." In the next chapter, we'll explore how stars are born, how most of them enter the main sequence, and then how they leave the main sequence and eventually die—either quietly, or with a cataclysm that ends up producing the exotic objects known as neutron stars and black holes.

THE LIVES AND DEATHS OF THE STARS

We know that stars must endure for periods of millions to billions of years. With human lives and even human history being so brief in comparison, how can we possibly determine what happens during the entire course of a star's life? How can we know what a star is like at birth, in infancy, through "adulthood," at old age, and eventually at death?

The answer, of course, is that we see in space countless examples of what must be stars at the different stages of life and death. If we can study many of these individual "still images" from the "movie" of stellar life, we should be able to understand how stars are born, evolve, and die.

Our first step is to determine the processes by which stars shine. If we can figure this out, we should be able to understand the nature of stars at each stage of their lives and deaths.

How Stars "Burn" Hydrogen

Hydrogen and helium are the two simplest elements in the universe, the ones with the fewest protons, electrons, and neutrons. So it should not be surprising to hear that spectral studies have found them to be by far the most common elements in the universe. That is not to say that they are common in a free form on Earth. In fact, helium was first identified in the nineteenth century in the spectrum of the Sun—the very term *helium* comes from the Greek *helios*, meaning "sun." But in outer space we know that hydrogen, the most common element, is the primary fuel that fires the stars.

We're speaking only figuratively here of "fire" and "fuel," though: the Sun and other stars do not shine by virtue of ordinary combustion of hydrogen or any other substance. If the Sun were powered in such a way, even its enormous mass would be consumed in a few thousand years. Instead, we know from many lines of evidence, including geologic ones, that the Sun must have been producing light and heat at not greatly unlike its current rate for roughly the past 4½ billion years. The only source of energy that could power the Sun and other stars for so long is nuclear.

So far, nuclear power plants on Earth have been able to control and use only *fission*, an inefficient (and messy) reaction involving the breakdown of heavy, complex elements, uranium and plutonium, into simpler ones. A far more efficient reaction is *fusion*, which combines simpler elements to form more complex, heavier ones (while releasing energy in the process). The reactions that make the Sun and stars shine are ones of nuclear fusion.

The easiest of these reactions to initiate and maintain is one that combines hydrogen atoms to produce helium atoms. The several-step process that occurs in the Sun and most stars we see in our sky is called the *proton-proton chain*. It can provide millions of years of stable energy production for even the most massive stars and billions of years for those whose mass is similar to that of the Sun.

But, of course, the process cannot continue forever, for eventually a star's supply of hydrogen begins to run short (this actually happens sooner, not later, with more massive stars, for they are more luminous and use up their hydrogen at a faster rate). What, then, is the fate of a star? The best way to learn the answer—or answers, for it depends on the mass of a particular star—is to step back and see how astronomers figured it out themselves.

The Hertzsprung-Russell Diagram

Astronomers have gained great insights into the evolution and fates of stars by the use of a simple diagram—a graph that has proven to be as fruitful as any that scientists have ever constructed. This graph, whose creation is attributed to Ejnar Hertzsprung and Henry Norris Russell, is called the *Hertzsprung-Russell diagram*, or **H-R diagram**.

The H-R diagram does not just display information that is already known. It points out new information, revealing patterns that are important clues to understanding the nature and evolution of stars.

Spectral types are plotted on the horizontal axis of the graph (since colors and surface temperatures correspond rather closely to spectral types, these, too, may be put on the horizontal axis). The vertical axis is true brightness—either in terms of luminosity or absolute magnitude.

The figure on page 73 shows an example of an H-R diagram. The first outstanding thing you notice is that many stars fall along a diagonal band running from upper left to lower right across the diagram. These are the stars of the *main sequence*, and the Sun is one of them. The main sequence stars are all purely hydrogen-burning. But that is not the case for the stars that are located in patches to the upper right and the lower left of the main sequence.

Far to the upper right of the Sun's position on the H-R diagram are many of the giant, bright giant, and supergiant stars that are not only far more

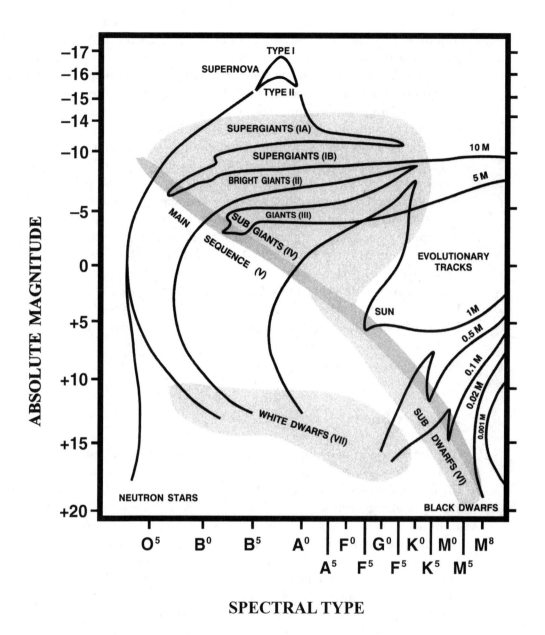

A Hertzsprung-Russell diagram. This H-R diagram plots spectral type versus absolute magnitude (true brightness). Various kinds of stars occupy different locations on the diagram, among them stars of the major "luminosity classes" (Ia and Ib, II, III, IV, V, VI, and VII). The Sun and most stars spend most of their life somewhere on the strip of the diagram that stretches from upper left to lower right—the main sequence (luminosity class V). The "evolutionary tracks" on this diagram are the life paths of stars of different masses (M = mass of the Sun). Massive stars burn brighter and go through their lives faster.

luminous and big than the Sun, but also much cooler and redder—red giants. Far to the lower left of the Sun's position on the H-R diagram are luminosity class VII stars that are not just less luminous and far smaller than the Sun but also (in most cases) much hotter and whiter—the white dwarfs.

The exciting realization that astronomers came to was that main sequence stars like the Sun turn, later in their lives, into red giants and then (unless they are very massive) into white dwarfs—which after many billions of years fade away, ending their lives on the lower left part of the diagram.

For of course the H-R diagram is typically drawn for the present era of time. But as millions and billions of years pass, stars evolve and change their positions, many of them, like the Sun, shooting from the main sequence to the upper right (the red giant region of the diagram) and then curving round to end up in the lower left (the white dwarf region of the diagram).

The path of a star on the H-R diagram—what happens in its life—depends principally on its mass. Later in this chapter, we will trace, with the help of the H-R diagram, the lives of three different kinds of stars: a star of low mass, a star of solar (moderate) mass, and a star of high mass. First, however, we will look at the one phase of a star's life that is fairly similar for all stars (just as it is for all people): its birth.

A Star Is Born

Stars condense from clouds of gas, primarily hydrogen, and especially in the concentrations of such gases we call nebulae. As a galaxy rotates, it may produce "density waves" that pass through the gas in its spiral arms. This could be a special trigger to setting off major outbreaks of star births. But, whatever movements make parts of a cloud of hydrogen gas dense enough, the fact remains that these regions do begin to contract under the force of their own increased gravity. As one of these *prestellar nebulae* contracts, it begins to rotate, for the energy of its previous random motions are conserved. It also begins to flatten and may eventually become a *proplyd*, a *protoplanetary disk* that will give rise to planets. And the energy of the inward-falling material produces heat—at the center of the prestellar nebula, a *protostar* begins to glow.

But it is not yet very hot. If there is not enough mass to keep the heating increasing, the object may become a *brown dwarf*, or, if even less massive, only a Jupiter-like planet (if such a planet is traveling through space alone or with other planets but with no star, it is called a *planemo*.) If there *is* enough mass, the protostar keeps warming, possibly beginning to shine out from open parts of what is now a *cocoon nebula*. And with similar neighboring cocoons it may be observed as a *Harbig-Haro object*. Leftover dark gas may appear against bright regions as black clumps known as *Bok globules.* As their glow brightens but is still only caused by the energy of gravitational contraction, the infant stars may vary in brightness and become *T Tauri stars.*

The protostar keeps contracting and heating up, moving leftward on the H-R diagram, reaching the main sequence at a height (luminosity) determined by its mass (the more massive, the higher on the H-R diagram). The object has achieved a great enough central pressure and temperature—about 10 or 15 million K—for nuclear fusion to begin. Hydrogen finally starts to be converted to helium. The outward-pushing pressure from the fusion process meets the inward-pulling force of gravity and finds equilibrium. A star is born.

The Life and Death of a Solar-Mass Star

After perhaps a billion years of going from nebula to hydrogen-burning, a star of solar mass now lingers on the main sequence for perhaps 10 billion years. But let's drop the dry term "solar-mass star." Let's instead talk about the life and fate of the solar-mass star that we all care deeply about—the Sun.

Our own Sun has been burning for about 4½ billion years and has generally been thought to be about halfway through its tenure of stable life on the main sequence. Some recent theories suggest that we might have much less time before changes in the Sun begin to cause critical problems for life on Earth—maybe just a paltry 1 billion years!

Although we may not know exactly how many billions of years from now the Sun will start to change critically, we do know that eventually this must happen. There is, after all, a finite amount of hydrogen in the Sun. What happens when all the hydrogen in the Sun's core has been converted to helium?

The answer is that gravity will begin to take over again (as it did with the proto-Sun), and the core of the Sun will start to contract. The rest of the Sun will still contain hydrogen, so fusion of that hydrogen will continue on in a shell around the core. In fact, the contraction of the core will produce greater heat, and this heat will prompt the hydrogen in the shell to fuse at a faster rate. What will happen is that the outer layers of the Sun will begin to expand. The total amount of light and energy coming from the Sun will greatly increase while the outer layers will expand so hugely that the surface will decrease in temperature. The Sun will swell and brighten but its surface will cool and therefore redden. It will move upward but also rightward on the H-R diagram. It will become a red giant.

Just how large the Sun will eventually become is not known. It will never grow to the behemoth size of current red supergiants like Betelgeuse and

Antares, for those stars are many times more massive than the Sun. But the Sun will surely grow to engulf the orbits of Mercury and Venus and perhaps swell out as far as the Earth. By that time, life on Earth will, in any case, be boiled away (unless the human race—or some other agency—has moved the planet).

But let's switch to the present tense to follow these momentous future events.

After reaching its maximum red-giant size, the core of the Sun continues to collapse. Eventually it reaches a density and temperature at which a new fusion reaction can occur: fusing helium to create carbon.

The helium creates greater energy but also is used up more quickly. The Sun—or any star in its approximate mass-range—keeps generating a higher density and pressure as its core contracts and, in turn, fuses heavier and heavier elements. Throughout this period, a star is alternately swelling and shrinking to try to maintain a balance. The instability is what makes such stars—our Sun in the future and any star in this state now—fluctuate in brightness. In other words, this is what causes a star to become what we call a variable star.

The heaviest element that steady fusion can produce is iron. But the Sun is not massive enough to reach that stage. Its final pulsations cough off some of its mass, relatively gently, to become a planetary nebula. Fusion ceases, and without its outward pressure, the star collapses into a body roughly the size of Earth. The contraction stops when the star has been crushed into a substance in which the electrons of its atoms have been stripped and mingled but still cushion the atomic nuclei from one another. We call it *degenerate matter*. The energy from contraction has made the star white-hot. Its luminosity has decreased greatly, but the temperature has soared: so on the H-R diagram, the star—our Sun in the future—has zoomed far leftward and far downward. It has become a **white dwarf**.

The planetary nebula of the white-dwarf Sun (or other white-dwarf star) remains energized and glowing for many thousands of years before the nebula dissipates. Then the white-dwarf Sun continues to radiate its remaining great heat off for many billions of years—until it eventually fades out and becomes a dark remnant, a *black dwarf*.

The Life and Death of a Low-Mass and a High-Mass Star

The story of the Sun and stars of similar mass is dramatic enough. But what happens to stars of much lower and higher masses? The life and death of the former is rather quiet. In contrast, the fate of a high-mass star is far more spectacular than that of our Sun.

First, let's consider the low-mass star. It begins life by contracting more slowly than the Sun did, taking much longer to reach the main sequence and begin hydrogen fusion.

If about half the mass of the Sun, a star will probably never be able to fuse any elements heavier than helium, never become a red giant, and never lose a substantial amount of mass. It will, after vast periods of time, become a white dwarf, but not as hot as more massive stars get to be. Then, like the Sun, its white dwarf slowly fades.

But suppose we start out with a star of only about one-tenth the mass of the Sun. Such an object will never get enough pressure and temperature at its center to go beyond fusing hydrogen into helium. It will burn hydrogen slowly, dimly, for many billions of light-years as a low-luminosity, low-temperature **red dwarf**. It will never become a white dwarf. It will simply exhaust its hydrogen and fade out.

Red dwarfs are tremendously more common than other kinds of stars, partly because of their long lifetimes. Indeed, astronomers believe that when the rest of the universe's stars have all faded out, there will eventually be only red dwarfs left.

Now let's switch from the slow, quiet, mostly uneventful life of red dwarfs to the opposite extreme—the rapid, tumultuous, and ultimately catastrophic life of stars much more massive than the Sun.

Let's consider the life of a star with a hefty ten times the mass of the Sun. Being so massive, the star will also be extremely luminous and will use up the fuel of itself at a prodigal rate. A star like the Sun may last 10 billion years before becoming a red giant and then take another few hundred million years before becoming a white dwarf. But a star ten times as massive as the Sun might live only 10 million years before destroying itself in a cataclysmic event that takes less than a second!

Each stage of a massive star's life is faster. It contracts and heats rapidly, reaching the main sequence at the upper left end of the H-R diagram—extremely luminous and extremely hot. Some of these stars would get hot enough to reach O spectral class, the hottest and rarest of the major types. For every one of these stars, there are thousands of stars of Sun-like mass and hundreds of thousands of stars of red-dwarf mass in our galaxy.

The massive star rapidly goes through fusing heavier and heavier elements, generating shell within shell inside itself. It swells, and its surface cools to become a red giant, but so complex and vigorous are its transitions into different fusions that it may pass rapidly back and forth through the range of spectral types (although all the while maintaining its great luminosity). Finally, all the elements up to iron have been fused in the core. But iron requires too much pressure and temperature to fuse. The powerful gravity of so massive an object takes over, causing a collapse of the core so sudden and violent—truly an implosion—that it sets off an explosion in the rest of the star. The explosion is the mighty event we call a supernova. It will send most of the mass of the star flying outward at speeds of millions of miles an hour. The star can for a few days shine as bright as 10 billion Suns, as much as the light of a medium-size galaxy! And yet ten times more power goes into the kinetic energy of the outflying material and a hundred times as much is carried off by subatomic particles called *neutrinos*. The glowing and growing cloud of ejected material may shine for a few years or for thousands as an SNR—a supernova remnant.

But the material strewn through space by a supernova is not merely beautiful to look at while it continues to glow. It is vital to spreading throughout space the heavier elements like carbon, which are essential to life. Supernovae may be the only mechanism that can create these heavier elements. Without them, there would be no Population I stars, no new generations of stars born from the material of old. (Our Sun's chemical abundances suggest that it is a third-generation star.) Without supernovae, there would be no planets possessing the heavier elements and therefore no possibility of life developing. Human beings are made of material that was created in supernovae. As Carl Sagan first said, "We are all star-stuff."

The composition of a star determines which of several varieties of high-mass supernova occurs (remember from the previous chapter that there *is* one kind of relatively low-mass supernova that can occur—when a white dwarf has too much material fall on it from a very close companion star). But what is of most fundamental importance is not the chemical composition that leads to different varieties of high-mass supernovae. What is important—at least insofar as what the supernova leaves—is how massive the star is. For that will determine whether the core collapses to become a neutron star or a black hole.

Neutron Stars and Black Holes

We saw that in a star of modest mass, its ultimate fate was to have its core collapse into an Earth-size white dwarf composed of degenerate matter. But if a star is massive enough (more than about eight solar-masses), it will go supernova and will leave a massive core that will collapse into something tremendously more compact. If the original star does not contain a lot more than eight solar masses, the core it leaves will not be a 10,000-mile-wide white dwarf but maybe only 10 miles across. It will be a **neutron star**.

Whereas in white dwarfs all the electrons are stripped and mingled but the atomic nuclei remain intact, now all protons and electrons are crushed together and become neutrons. (In this process, stupendous numbers of neutrinos are produced—the neutrinos in a supernova blast.) The neutron star is something like a city-size atomic nucleus, but made entirely of neutrons.

A further amazing fact is how fast a neutron star rotates. A newly formed neutron star can spin at a rate of hundreds of times per second (imagine a ball the size of a city doing that!). We have observational evidence of this because we can detect some neutron stars as **pulsars**. Back in the 1960s, astronomers first detected incredibly rapid and regular radio-wave pulses reaching Earth from point sources in distant space. There was shock, even some speculation that they were beacons designed by extraterrestrial intelligence. Instead, what we are detecting—usually at radio but sometimes also at other, even optical, wavelengths—is a beam of energy from neutron stars. Theorists tell us that a neutron star should indeed emit a beam of energy from each of its polar regions, where the energy escapes the neutron star's magnetic field. If the spin axis of a neutron star is pointed in our direction, the beam passes across us for an instant during each rapid rotation, registering on radio telescopes, electronic imagers, or other detection devices. Otherwise, we get no pulse. So only some neutron stars appear to us as pulsars.

Now suppose we have an extremely massive star—much more than eight times the Sun's mass. It shines for a few million years, then goes supernova. But the force of its gravity is so great that its core's collapse cannot be stopped even by "the strong interatomic force." Even neutrons collapse into one another to become a "singularity"—the perhaps infinitely small object that is called a **black hole**. A black hole is an object whose gravity is so strong that nothing, not even light or other forms of electromagnetic radiation, can escape from it. It can contain the mass of a single star or, as is the case in the hearts of many galaxies, can keep on devouring star systems until its mass is millions or even billions of times greater than our Sun. No stranger or more imposing phenomenon is known to exist in nature.

PART THREE

PROFILES OF THE BRIGHTEST STARS

SIRIUS

8

Sirius is the Dog Star, the brightest star of Canis Major, the Great Dog and the brightest of all stars in the night sky. But Sirius is much more than that. Sirius is the star of stars, the king of stars, the star par excellence. Discovering Sirius for myself out my bedroom window at the tender age of six probably did more than any other single event to inspire my lifelong love of astronomy. That first experience defined for me forever the beauty and wonder of a star.

What makes Sirius marvelous other than its extraordinary brightness?

- The placement of Sirius: atop bright Canis Major, paired with the other Dog Star Procyon, pointed to by Orion's Belt.

- The color of Sirius: in one way (as we'll see), Sirius is more prominently colorful than any other star.

- The interest of Sirius: it's the closest naked-eye star for most of the world's population, the bright star we're departing from in space, and the closest star to possess a white-dwarf companion.

- The lore of Sirius: it has been intimately connected with the life-giving annual flood of the Nile, with the origin of honey, with the summertime madness of dogs, with the fabulous bird called the phoenix.

Sirius is the only individual star to have been used as the basis for the calendar of one of history's greatest civilizations. It is the only star that has given rise to a phrase and phenomenon name that are common even today in Western languages.

Canis Major is one of the very brightest constellations. But as William Tyler Olcott wrote in a chapter on Canis Major in *Star Lore of All Ages*: "The importance of the constellation is overshadowed by the fame of its lucida [brightest star], Sirius the 'King of Suns,' concerning which star volumes have been written. Its matchless brilliancy has inspired the poets of all ages, and historically Sirius is beyond question the most interesting of all the stars in the firmament."

Hubble Space Telescope image of Sirius A and B (lower left, near long diffraction spike).

The Brilliance of Sirius

In the midst of its many other marvelous attributes, Sirius is, first and foremost, *bright*.

The most reliable measurement of Sirius's visual magnitude seems to be −1.44. What does that figure mean? It means Sirius is about twice as bright as its closest rival (Canopus) and about three times as bright as its second-closest rival (Alpha Centauri). Those stars are not properly visible north of the tropics. But if you take the next four stars—the 2nd, 3rd, 4th, and 5th brightest visible from 40° N—and combine all four (Arcturus, Vega, Capella, Rigel), the total apparent brightness would be scarcely brighter than Sirius alone! Sirius is almost ten times brighter than a standard star of magnitude 1.0 (such as Spica) and about 1,600 times brighter than a star of magnitude 6.5 (the dimmest star easily visible to the average person observing with the naked eye in a clear and very dark sky).

I've seen Sirius with the naked eye just before sunset, but under excellent conditions an observer could surely detect it sooner. The nineteenth century deep-sky writer T. W. Webb says that Sirius was seen by Bond at Harvard in broad daylight with the naked eye and it has been observed with a telescope of as little as one-half inch aperture at noon. Since I know observers who have seen Jupiter with the naked eye many hours after sunrise when it wasn't all that much brighter than Sirius, I suspect the same could be done with the star. What about at night? According to one calculation, Sirius should be just bright enough to cast shadows on white surfaces (snow or a white sheet spread on the ground) in near-ideal dark conditions.

What is brighter than Sirius? Two planets are, when visible, always brighter: Venus and Jupiter. But Venus can never be seen high in the middle of the night. And neither Venus nor Jupiter can twinkle like Sirius—something that, as we'll see in a moment, makes a tremendous difference in beauty.

Has any star ever outshined Sirius in our sky? In the past 90,000 years, only a few novae and supernovae have done so—briefly. Actually, the brightest nova on record, one in Aquila in 1918, fell just short of equaling Sirius's brightness. A few supernovae in the past 2,000 years have greatly exceeded the brightness of Sirius—but only for a while. By the way, to get some kind of idea of what the brightest recorded supernova, the approximately magnitude $-7\frac{1}{2}$ Lupus supernova of 1006, looked like at peak to the naked eye, you can train a common 4- or 6-inch telescope on Sirius.

Quite apart from any comparisons to stupendously rare supernovae, the brightness of Sirius through a sizable telescope is simply astounding. Astronomy writer Robert Burnham Jr. says that for the great eighteenth- and nineteenth-century astronomers the Herschels, the approach of Sirius to the field of view in one of their giant reflectors "was heralded by a glow resembling a coming dawn, and its actual entrance was almost intolerable to the eye." He may be referring in part to what William Herschel said about Sirius in the field of his mighty 48-inch reflector (the largest in the world at the time). Sir William said that Sirius entered his view "with all the splendor of the rising Sun, and forced me to take my eye away from that beautiful sight."

The Scintillation and True Color of Sirius

What makes the view of Sirius in a telescope breathtaking is not just its brightness but, on many nights, the wild fluctuations in its intensity and its *colors*.

Yes, I said *colors*—the plural. And the reason I said it is because Sirius often *scintillates* with many colors. Nor do you need a telescope to see this phenomenon. In fact, it may be even more wondrous to look with the naked eye and see, in a landscape of dark trees and skyscape of many stars, Sirius sparkling and darting from its heart rays and bursts of every color you can imagine in unpredictable sequence of hue and intensity.

What is scintillation? It is the technical term for "twinkling." Scintillation occurs when cells of turbulence in Earth's atmosphere cause the light coming to us from a star to deviate from a straight path. Colors are different wavelengths of light, and different wavelengths are deviated by different amounts by Earth's unsteady atmosphere. So the mostly white (slightly blue) light of Sirius can get broken into this display of prismatic, many-hued splendor. And a remarkable fact is that only Sirius is bright enough for its scintillation to display prominent colors to the naked eye frequently. An additional factor that often makes this phenomenon noticeable to observers at mid-northern latitudes is the relatively low arc the star takes across the sky at those latitudes. The lower in the sky a star appears, the longer the pathway of Earth's atmosphere its light has to pass through—and the more cells of turbulence. Watch especially for the rainbow colors of Sirius in the first hour or two after the star rises, before it gets higher and shines more steadily. In her book *The Friendly Stars*, astronomy writer Martha Evans Martin described the experience of following Sirius up the sky until its light steadies more and the prismatic colors settle out: "He [Sirius] comes richly dight in many colours, twinkling fast, and changing with each motion from tints of ruby to sapphire and emerald and amethyst. As he rises higher and higher in the sky he gains composure and his beams now sparkle like the most brilliant diamond, not pure white but slightly tinged with iridescence."

Another famous quote about the color changes of Sirius comes from Alfred, Lord Tennyson in *The Princess*. The poet writes:

> The fiery Sirius alters hue
> And *bickers* into red and emerald.

The italics are mine; the exquisite word choice Tennyson's—and the beauty Sirius's.

The color changes of Sirius are noticed by even casual skywatchers. The early nineteenth-century French astronomer Dominique François Arago said that as many as thirty color changes in a second had been observed in Sirius. According to R. H. Allen in *Star Names: Their Lore and Meanings*, Arago also "gave *Barakish* as an Arabic designation for Sirius, meaning Of a Thousand Colors."

As captivating as Sirius's dance of many colors is, there is great satisfaction in seeing and ascertaining the star's true hue. What is the color of Sirius when it is little disturbed by atmospheric turbulence? Some people say white—as described in Martin's passage. But I've always seen a distinct tint of blue in Sirius. That's aesthetically appropriate for a winter star, but I'm not being misled by any wishful thinking. Many famous observers have reported at least a hint of blue in Sirius and a few informal tests I've made with novice observers seem to show that most people do see this blue. Two thousand years ago, astronomy writer Manilius wrote of Sirius's "sky-blue face."

The Heliacal Rising of Sirius

When I watch, around August 12, the peak of the Perseid meteor shower in years when weather permits, Orion rises just as morning twilight is nearing. The dawn then always brightens just a little too soon for me to spot Sirius coming up. A few days later, however, low in the bright dawn, the great star makes its first showing—an event known as the *heliacal rising of Sirius.*

I see it in mid-August, a few weeks after a separate event, the rising of Sirius at the same time as the Sun from my latitude of about 40° N. From the latitude of Cairo, Egypt (about 30° N), Sirius rises at the same time as the Sun about one week earlier, so observers there might see the heliacal rising—the first dawn visibility of Sirius—around August 7. The exact date in a given year depends, of course, on sky conditions and the perception of the individual observer. But why this observation is especially interesting now is that it was once monumentally important—in ancient Egypt. For the Egyptians noticed that the heliacal rising of Sirius, as seen from their land, came at around the time of the summer solstice. And this coincidence made Sirius the great predictor of the natural event upon which the very life of Egyptian civilization depended in their desert land—the annual flood of the Nile.

Only this flood could make fertile the area near the Nile; it was alone what permitted the flourishing of the most famed of history's first civilizations. And the heliacal rising of Sirius was the flood's herald. No wonder the Egyptians decided to designate this rising as the day of their New Year. They counted 365 days until the next heliacal rising of Sirius and made this the period of their civil calendar.

Unfortunately, the true length of the year is close to 365¼ days. Without the use of a leap day, the Egyptian calendar advanced by a day every four years in relation to the seasons—and in relation to other natural phenomena, including the heliacal rising of Sirius and the flood of the Nile. So,

The heliacal rising of Sirius, as seen from ancient Egypt. Sirius is the white dot just above the boat's right end. The constellations corresponding to our Orion, Taurus, and Gemini (or Gemini's feet) are pictured. Note the vertical line of three stars, Orion's Belt, just left of the archaic long ear of its observer—one of the first domesticated cats.

although the heliacal rising still announced the flood, the date of these events fell later and later in the calendar. After a period of 1,460 years (365 divided by ¼ = 1,460), the date progressed all the way through the year and came back to its original point.

One name for Sirius in ancient Egypt was Sihor—"the Nile Star" (also "the Fair Star of the Waters"). But an even more important name for Sirius was Sopdet, the "preparer" or "preceder"—of the Nile flood. Sopdet became "Sothis" in Greek, and so the 1,460-year period became known as *the Sothic cycle.*

This length of time was also referred to as "the Period of the Phoenix," for that fabulous bird is rumored to live 1,460 years and then perish in fire— only to be born again from its own ashes.

Thanks to one person, the Roman historian Censorinus, we know that Sothic cycles must have begun (each with the coincidence of the heliacal rising and the Nile flood) in 4242 BC, 2782 BC, and 1322 BC. Which of these dates was the one at which the connection between the heliacal rising and the flood was first noticed and, supposedly, the calendar introduced? Not all that many decades ago, historians thought that the Egyptian civilization got its start before 4000 BC. We now believe that not until about 3500 BC was there a "unification of Nile Valley polities," and not until more like 3100 BC or later did the dynasties and the Old Kingdom begin. So the launching of the civil calendar with the heliacal rising as New Year may have been instituted around 2782 BC (probably a few hundred years before the building of the Great Pyramid of Giza).

Today, the Nile flood can't reach Memphis, the ancient capital of Egypt (near present-day Cairo)—for it is restrained by the great Aswan Dam. But even if it could, the heliacal rising of Sirius wouldn't be there to greet it. For the flood is caused by meteorological conditions farther up the river that still send a surge down the river around the time of the summer solstice. But the slow change in Earth's axis direction called precession (see chapter 3) has advanced Sirius (and all other stars) in relation to Earth's seasons. Precession is extremely slow but, in the thousands of years since the establishment of the Egyptian civil calendar, it has moved the heliacal rising of Sirius to later in the summer at the latitude of Egypt.

The civilization of ancient Egypt itself endured for several thousand years, and across that long span Sirius became identified with several different important gods and goddesses of the Egyptians. One was Anubis, the jackal-headed god who led the dead into the hall of judgment and weighed their hearts in a balance against the feather of truth. Another was ibis-headed Thoth. Yet another was the goddess Hathor, represented as a cow or cow-headed woman. The former is the case in the famous carving of constellations—including the zodiac—that is found on the ceiling of the temple of Hathor at Denderah in Egypt. The cow representing Hathor has a star between its horns that is clearly intended to be Sirius.

Another legend tells that the soul of the goddess Isis finally found peace residing in Sirius.

The Name Sirius and the Dog Days

Sirius is mentioned in the writings of Homer and Hesiod, but they called it Kyon, "the Dog." Indeed, later, the eighth brightest star in the heavens, Procyon, received its name, which means "before the Dog," in reference to its rising just a little before Sirius for viewers at midnorthern latitudes. So mighty is Sirius that even another star's name is derived from that star's role of coming up just before Sirius: magnitude 1.98 Beta Canis Majoris is called Murzim or Mirzam in Arabic, the "roarer" that announces the coming of Sirius.

Today's amateur astronomer will soon grow weary of hearing jokes based on the fact that the name Sirius is pronounced like the English word *serious*. By the way, although it is tempting to think of Sirius being "searingly" bright, the verb *sear* is from Old English and so presumably not related to the name of the star. But the name of Sirius does in part have something to do with burning.

That name—Sirius—is one of the three oldest star names in use by modern Western culture. (The other two that also seem to date from prescientific Greek times are Arcturus and Canopus.) The Greek form of Sirius is Seirios, which is usually translated as meaning "sparkling" or "scorching," references to its light and imagined heat. Consider what Eratosthenes writes about the use of *seirios* as an adjective for stars: "Such stars astronomers call *seirios* on account of the tremulous motion of their light." On the other hand, Aratus wrote:

> In his [Canis Major's] fell jaw
> Flames a star above all others with searing beams
> Fiercely burning, called by mortals Sirius.

This is just one of innumerable references not only to the supposed heat of Sirius but to its combining that heat in summer with the Sun to produce the "dog days" (also called "canicular days").

Skywatchers knew that in the hottest weeks of the year, after the summer solstice, Sirius was traveling the day sky (though unseen in the bright sky) with the Sun. But the exact range of dates of the dog days has always been debatable.

The Old Farmer's Almanac says it follows a tradition that the dog days last the forty days from July 3 to August 11, the latter date being that of the heliacal rising of Sirius (as we've seen, attributing a specific date to that rising is not very meaningful if we don't at least specify the latitude of the observer).

One can compile a long list of classic Greek and Roman writers who mention the heat of Sirius and the accompanying ill effects it produces when added to the Sun's during the dog days. According to R. H. Allen, Hippocrates (the "father of medicine") "made much, in his *Epidemics* and *Aphorisms*, of this star's power over the weather, and the consequent physical effect upon mankind." Yet, much later in history, Dante spoke of "the great scourge of days canicular," and Milton in *Lycidas* called Sirius "the swart [malignant] star."

Homer didn't really emphasize the heat of Sirius but did attribute to it a maleficent effect. In Lord Derby's nineteenth-century translation of Homer we read of Sirius:

> The brightest he, but sign to mortal man
> Of evil augury.

But earlier, Alexander Pope's translation of the same lines took extreme liberties, adding reference to the supposed heat and alleged pestilential effect of Sirius:

> Terrific glory; for his burning breath
> Taints the red air with fevers, plagues and death

Homer's most famous references to Sirius are when he compares it to Diomedes' shield and to Achilles. Homer refers to Sirius as "the Autumn Star" but, according to R. H. Allen, "The season intended was the last days of July, all of August, and part of September—the latter part of summer." Lord Derby's translation of the comparison of Sirius to the shield of Diomedes says:

> A fiery light
> There flash'd, like autumn's star, that brightest shines
> When newly risen from his ocean bath.

And when Priam sees Achilles, he compares the mighty warrior to

> Th'autumnal star, whose brilliant ray
> Shines eminent amid the depth of night,
> Whom men the dog-star of Orion call.

There is much more lore and tradition to be told about Sirius, from ancient Greek times right up to the twenty-first century. We'll return to it near the end of this chapter. But now it is time to switch to a consideration of the fascinating nature of Sirius as a sun—or, rather, as two.

Sirius and Sirius B as Suns

Two thousand years ago, Manilius wrote of Sirius that "hardly is it inferior to the sun, save that its abode is far away." He was right, but compared to most stars Sirius is close to us. Sirius lies 8.60 light-years from Earth, making it the seventh closest star system known—but the second closest visible to the naked eye. Sirius is almost exactly twice as far away as the Alpha Centauri system, but the latter is too far south to be seen by much of the world's population. Sirius is over half a million times farther from Earth than our Sun is.

It's always interesting when stargazing with beginners to match people of different ages with a star whose light left it around the time they were born. In the case of Sirius, children who are about eight or nine can be told that the light they are seeing from Sirius tonight departed from it when they were born. The journey of Sirius-light is as old as they.

We must not suppose that Sirius is only bright because it is close, however. There are more than one hundred stars within 20 light-years of Earth. Sirius is the most luminous star we encounter until we meet Vega, at a distance of 25 light-years from Earth. Sirius is several times more luminous than its brightest competitors, Altair and Procyon, within this distance of us. The absolute magnitude of Sirius is +1.45 (so if it were 32.6 light-years away from us, it would still just make the cut as a 1st-magnitude star).

But let's talk about the nature of Sirius as a sun compared to our own Sun. Sirius has 26 times the Sun's luminosity, about 1.75 times the Sun's diameter, and a little more than 2 times the Sun's mass. It is a main-sequence hydrogen-fusing dwarf (luminosity class V) star like the Sun, but its spectral type is A1—so its surface temperature is about 9,880 K, much hotter than the Sun's. Sirius rotates in less than 5.5 days (at least five times faster than the Sun), with a minimum equatorial speed of 16 kilometers per second. For some reason, Sirius's spectrum shows it to be "metal-rich," its iron content about three times that of the Sun.

Sirius is clearly a very interesting sun. But Sirius is not alone. It goes through space with a white-dwarf companion, known as Sirius B or sometimes, playfully, as "the Pup."

Sirius B is easily the most famous of all white dwarfs—it was the first discovered, and the first understood for what it is (we'll discuss that history

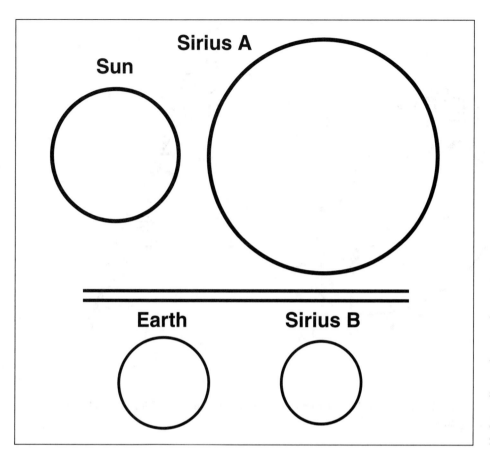

Sun

Sirius A

Earth

Sirius B

Relative sizes of the Sun compared to Sirius A and (on a different scale— Earth is less than one-hundredth as wide as the Sun) Earth compared to Sirius B.

more in a moment). Sirius B is the brightest of all white dwarfs in apparent magnitude because it is the closest. The data on the Pup leaves no doubt as to the nature of the star. Consider: Sirius B is hotter than Sirius, but the Pup is almost exactly 10 magnitudes—that is, 10,000 times—less bright than Sirius. (The Pup's luminosity is less than $\frac{1}{400}$ that of the Sun or, even if we include its large amount of ultraviolet radiation, only about 2.4 percent of the Sun's—but in X-ray imagery it vastly outshines even Sirius). Its weak light output despite a very hot surface can only mean that Sirius B is tremendously smaller than Sirius or, for that matter, than the Sun. In fact, calculations show that Sirius B must have only about 0.92 the diameter of *Earth*. Yet studies of the motions of Sirius and Sirius B through space demonstrate that the latter has almost exactly the same mass as the Sun—1.00 plus or minus 0.02 solar masses (almost exactly half that of Sirius A). We can only conclude that Sirius B is tens of thousands of times denser than the Sun. Its "degenerate matter" would weigh several tons per cubic inch.

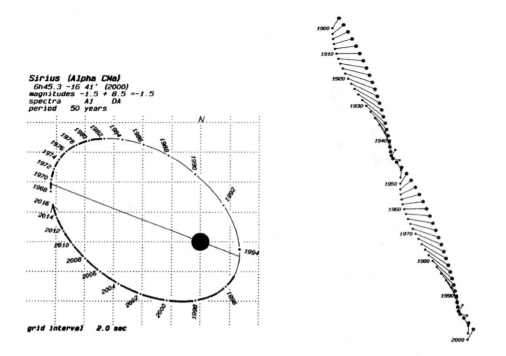

The orbit of Sirius B around Sirius A (left diagram) and the "braided" motion of Sirius A and B across the sky (right diagram).

The average separation between Sirius and its companion is 19.8 AU (about the distance between the Sun and Uranus) but varies from 8.1 AU (less than the Sun-to-Saturn distance) to 31.5 AU (about the Sun-to-Neptune distance). The two stars revolve around a barycenter that is closer to the more massive star (Sirius), taking 50.1 years to complete one orbit. The figure on the right, above, shows the beautiful entwining of the two stars' paths through space into a kind of braid. The last *periastron* (closest-together orbital positions of the two stars) was in 1994. The last *apastron* (farthest-apart orbital positions of the two stars) was in 1969 (the next is in 2019). Periastron and apastron are not when Sirius and Sirius B *appear* closest together and farthest apart in the sky, however, for we are not seeing the orbit face-on. Instead, the orbit is tilted toward us at about a 44° angle, with Sirius B currently on the near side of the orbit to us, closer than Sirius (see the figure on the left, above). The separation of the Dog Star and the Pup in the sky was greatest in 1974 (11"), least in 1993 (2.5"), and will be greatest again in 2022.

Discovering and Observing Sirius B

Observing the companion of Sirius through an amateur telescope is a challenge even when the two stars are farther apart than usual (as will be the case for the next few decades). The story of the Pup's prediction and discovery is an exciting tale.

You will recall that in 1718 Halley announced his discovery of proper motion that he had made by comparing the positions of Sirius, Arcturus, and Aldebaran in his time to those recorded in ancient times. In the years from 1834 to 1844, the mathematician and astronomer F. W. Bessel made a detailed study of the proper motion of Sirius and discovered that it wasn't straight. He came to the conclusion that this deviation was caused by the gravitational pull of a yet-unseen companion star that orbited Sirius in a period of about fifty years. In 1851, astronomer C. H. F. Peters computed a theoretical orbit for the companion, but many searches by skilled observers failed to locate the object. Finally, in January 1862, Sirius B was first sighted, as part of the testing of optics expert Alvan G. Clark's superb new 18½-inch refractor by his son. Bond soon confirmed the observation, using the 15-inch refractor at Harvard Observatory.

The glare of Sirius prevented determination of Sirius B's spectral type and surface temperature until W. S. Adams's work in 1915. Only then was it confirmed that the Pup must be the strange kind of star we call a white dwarf.

How easy is it for amateur astronomers to detect the companion of Sirius? In years when the pair appear close, only excellent seeing and a very large telescope will avail. But in years when the two stars are well separated, a skilled observer may detect Sirius B through a much smaller aperture. The observer must have good seeing (try the observation when Sirius is near its highest) and use clean optics. Furthermore, if the telescope uses a "spider" (the vaned diagonal-mirror holder in many reflectors), the observer must make sure Sirius B is not hidden behind one of the diffraction spikes of light that a spider produces. Two traditional warm-ups for the feat are trying to spot the companions of Rigel and Adhara. If these can't be achieved, there is no chance of making the much more difficult sighting of the Pup.

In the winter of 1962, with Sirius and Sirius B more than their average distance apart, Robert Burnham Jr. conducted tests of the latter's visibility with the 24-inch Clark refractor at the Lowell Observatory. These showed that Sirius B was most easily seen with the telescope stopped down to 18 inches but was just detectable with the aperture stopped down to as little as 6 inches (this was very difficult and only possible because he already knew exactly where to look). Burnham used magnifications ranging from 200× to 900×. With the higher powers in this range (which our unsteady atmosphere rarely lets us use even with very large telescopes), Burnham was able to view the Pup with the Dog Star itself entirely outside the field!

Burnham says that on many subsequent occasions he was able to observe the companion star with a 10-inch. And I know of several observers (Walter Scott Houston and Steve Albers) who have detected Sirius B with high-quality 6-inch telescopes. But such observations can be made only in years when the two stars are fairly well separated. Houston seems to suggest that the Dog Star and the Pup need to be about 5" apart before the latter can be seen in amateur telescopes. Fortunately, the separation is now growing. During 2006, it increased from 7.19" to 7.62", and reports of Pup sightings began to be heard. The gap will keep widening until it reaches 11" in 2022.

Two of the most interesting observations of the companion to Sirius I've ever heard about were reported to me in 2005 by the New Hampshire amateur astronomer Albert Doolittle. Doolittle says that in the late 1970s he and some other members of the Amateur Telescope Makers of Boston got a rare chance to use the 15-inch refractor at Harvard Observatory. He got an opportunity to view Sirius B in this telescope, the very one that Bond had used in 1862 to confirm Clark's discovery of the Pup. Then, in March 1980, Doolittle was visiting Tanzania to observe a total eclipse of the Sun. While there, he spent a night on the Serengeti Plain, observing with a 3½-inch Questar telescope. Sirius was almost overhead at that latitude, and he was clearly able to see the companion with this extremely small—but excellent—telescope! He was watching the king of stars and its hidden companion straight above in the African darkness—while hearing an occasional lion's roar in the distance.

The Lives of Sirius and Sirius B

If you think that Sirius and Sirius B make an interesting pair now, you should have seen them in the past. Most experts estimate the age of these stars at 225 to 250 million years. That's only about one-twentieth as old as the Sun. It's only just long enough for them to have completed one orbit around the galaxy. The remarkable thing to know is that for about half of their lives to date, Sirius B was much more massive and large than Sirius. To have its current mass as a white dwarf, Sirius B must have started out as a hot blue star of spectral type B, several times more luminous than today's Regulus. It would have been about five times more massive than the Sun. But because of that greater mass, what we now call Sirius B evolved quickly and swelled to become a red giant. The red giant lost about 80 percent of its mass (relatively peacefully) when it turned into a white dwarf—about 124 million years ago. It never got quite big enough as a red giant for its material to interact with and disrupt Sirius, and presumably if it produced a planetary nebula, that, too, did not much disturb Sirius.

What a sight this star system would have been if it had passed its current distance from Earth in its early days. The point of light would have shined brighter than Venus in our skies, bright enough to cast shadows.

Of course, someday Sirius itself will go red giant and then become a white dwarf. This won't happen for hundreds of millions of years; the lifetime of an A-type star before it becomes a white dwarf is about a billion years.

Red Sirius, the Dogon, and Triple Sirius

Is it possible that Sirius B became a white dwarf as recently as two thousand years ago? A number of ancient writers seem to say that Sirius was reddish (others, such as Manilius, speak clearly of its blueness). In, *Burnham's Celestial Handbook*, Robert Burnham lists Aratus, Cicero, Horace, Ptolemy, Seneca, and others as having made comments indicating the Dog Star's ruddiness. Could Sirius B have still been a red giant then, dominating the coloration of the point of light called Sirius even to the naked eye?

Experts on stellar evolution say no, that the transition from red giant to white dwarf takes hundreds of thousands of years, not two thousand. And as we just discussed, the latest research suggests that the change occurred over a hundred million years ago. Some of the ancient descriptions may suffer from poor translations: fiery, for instance, does not necessarily mean "red." But presumably most of the comments are merely references to red produced in the scintillations of Sirius when it is low or when the atmosphere is especially unsteady. It's true that red is not continuous in the scintillations and that the other colors appear also. But the red may be the most impressive and noteworthy. Seneca said that Sirius was not just as red but *redder* than Mars—which would mean redder than the red giant stars Betelgeuse and Antares, too. The only conceivable way that that could be true is if the pure pulses of red in the scintillation of Sirius were what Seneca was observing.

There is a much more outrageous controversy linked to Sirius B: that involving the Dogon tribe of Mali, who "build villages of beehive huts down the banks of the upper Niger" in Africa. When anthropologists made what they believed was the first visit ever of technologically advanced people to the Dogon, they were shocked to hear some of the Dogon stories about objects in the heavens. The Dogon seemed to know about the rings of Saturn, which of course are only really apparent through telescopes. Even more startlingly, the Dogon told the anthropologists that Sirius has a satellite (some accounts say two) made of *sogolu*—a metal so heavy that a grain of it weighs as much as 480 donkey-loads. This, of course, sounds like a technologically simple culture's description of the degenerate matter in a white dwarf!

What could be the explanation for the Dogon having such knowledge? Some sensationalistic fringe-science writers have suggested that the most likely explanation is that extraterrestrials visited ancient Egypt and Sumeria and the knowledge they imparted about the companion of Sirius passed down to the Dogon! A philosophic principle that scientists favor, called Occam's Razor, says that the simplest explanation that could be the correct one most often *is* the correct one. We don't have to assume extraterrestrials who made contact with ancient Earth civilizations, extraterrestrials who were interested in telling those civilizations details about the physical state of a companion of Sirius, extraterrestrials whose information about Sirius B to those civilizations was passed down for several thousand years and ended up being known by the Dogon—even though there is no record of the information in Egyptian or Sumerian records nor in the writings or lore of any people other than the Dogon. No, all we have to assume is that, unbeknownst to the anthropologists, the Dogon were visited by some earlier group or individual who had a telescope or at least knowledge about Sirius B (and Saturn), which that person shared with the tribe.

I mentioned before that some accounts have the Dogon claiming that Sirius has *two* satellites. Well, some highly skilled observers of Sirius B in the twentieth century thought that that they did see the point of Sirius B look double. Could there be a third member of the Sirius system? The increasingly powerful tools of astronomers have not in recent years been able to confirm that a third star exists.

Sirius in Lore and Culture before the Twentieth Century

A reasonably complete guide to the lore of the brightest stars would fill a long book of its own. In such a book Sirius might have the longest chapter. Here, I will touch upon only a brief selection of the lore beyond what we have already discussed.

An interesting point about Sirius is that it seems to have been identified as a dog or doglike creature in many ancient cultures. How did this characterization of the star get started? It has been suggested that Sirius was regarded as a faithful dog keeping guard to warn the Egyptians of the Nile flood. The Phoenicians, those greatest of early sailors, knew the star as Hannabeah, "the Barker." R.H. Allen mentions one authority who claimed that the Chinese identified Sirius as the "Heavenly Wolf," but whether this was in early Chinese history, and probably uninfluenced by the Egyptians, or occurred in much later times, he does not say.

Certainly some cultures had other names and imaginings for Sirius. The aboriginal people of Australia saw Sirius as an eagle. Burnham wrote that "the Arabic name Al Shi'ra resembles the Greek, Roman, and Egyptian names, and suggest a common origin from an older tongue, possibly Sanskrit, in which the name 'Surya,' the Sun god, simply means 'The Shining One.'" (Practitioners of yoga may recall that a famous series of postures is called *surya namaskar*, "the Sun salutation.") Sirius and the Sun, the brightest stars of night and day, were often connected in ancient times. A perhaps more definite Hindu title for Sirius, however, was "the Deerslayer."

The primacy of Sirius in Egyptian starlore is well known. Surprisingly less famous in modern times is Sirius's supposed role in the most long-lasting and renowned of pagan religious ceremonies, the Eleusinian mysteries. Allen says that "the culmination of this star [Sirius] at midnight was celebrated in the great temple of Ceres [the Greek goddess Demeter] at Eleusis, probably at the initiation of the Eleusinian mysteries." Ceres/Demeter was the goddess of agriculture and growing things, and the Eleusinian mysteries were supposedly held to welcome back the return of growth and greenness in the spring. Yet Sirius's midnight culmination (the reaching of its highest point in the south) now occurs at the opening minute of our New Year (midnight of January 1), and if the Eleusinian mysteries started before 1000 BC (which is what some authorities claim), the midnight culmination of Sirius would back then have occurred earlier than it does now—late autumn, not early spring.

The Greeks and Romans did not always regard Sirius as inauspicious and dangerous, causing the dog days and the maladies that went with them. Pliny credited the star as being the source of honey. As Allen quoted him in a quaint archaic translation: "This [honey] cometh from the ayer at the rysynge of certeyne starres, and especially at the rysynge of Sirius, and not before the rysynge of Vergiliae (which are the seven starres called Pleiades) in the sprynge of the day."

Allen says that Pliny nevertheless seemed to be in doubt as to whether honey was "the swette of heaven, or as it were a certeyne spettyl of the starres" and added that "this idea is first seen in Artistotle's *History of Animals*" (the idea that honey is the sweat of the heavens or the spittle of the stars—or both?). Allen says that latter-day astrologers have claimed that wealth and renown came to those born under Sirius and "its companion dog" (presumably Procyon). He concluded his passage on associations of Sirius with gentle and benevolent things by quoting a pair of lines from the poet Willis's work *The Scholar of Thebet ben Khorat*:

Mild Sirius tinct with dewy violet,
Set like a flower upon the breast of Eve.

Sirius may have additional significance to those people who live where the great star passes overhead. One such place is Lake Titicaca, high in the Andes (do we know of Inca legends about Sirius?). Another place where Sirius goes overhead is Tahiti. Polynesians who discovered the Hawaiian Islands where the star Arcturus was at the zenith were able to find their way home by knowing that Sirius was the Tahitian (and Samoan) zenith star. The remarkable Polynesian legend of Sirius, Aldebaran, the Pleiades, and the god Tane is discussed in our chapter on Aldebaran.

A profoundly different story is Voltaire's 1752 work "Micromegas." Featured characters are a being from a planet circling Sirius and his companion from Saturn. These two pay a visit to Earth, providing Voltaire with a vehicle for social commentary.

From a far-north land where Sirius shines quite low in the south comes a story about Sirius that is in one way the reverse of the Hervey Islanders' tale. Rather than telling about the shattering to pieces of a bright star, the nineteenth-century Finnish poet Zakris Topelius wrote that the great brightness of Sirius was produced by the merging of two glowing lovers, Zulamith the Bold and Salami the Fair. These two had been separated by the Milky Way but spent a thousand years building a bridge to cross it and reach each other. When the bridge was complete, they

Straight rushed into each other's arms
And melted into one;
So they became the brightest star
In heaven's high arch that dwelt—
Great Sirius, the mighty Sun
Beneath Orion's belt.

Sirius in Lore and Culture in the Twentieth and Twenty-First Centuries

The British philosopher Olaf Stapledon became famous for such influential twentieth-century science-fiction novels as *First and Last Men* (a fictional history of the human race over billions of years) and *Starmaker* (a fictional history of the entire universe). But he wrote more moving novels set on a less vast stage: *Odd John* (the story of a mutated superman) and *Sirius: A Fantasy of Love and Discord*—the story of a dog genetically engineered to have intelligence equal to that of humans. *Sirius* was published in 1944 and explores the dog's existential strivings to find meaning in life and a place in life to fit in.

In the twentieth century, J. R. R. Tolkien's various versions of his tale *The Silmarillion* mention the names of stars given to them by Tolkien's "Quendi" or Elves. Sirius is said to be called Helluin. We are told that the Elves first awoke at the very hour when Menelmakar (or Menelmacar)—Orion—"first strode up the sky and the blue fire of Helluin flickered in the mists above the borders of the world." Later, in Tolkien's book *Morgoth's Ring*, we read that the jewels devised by the greatest of Elvish craftsmen, Fëanor, were at first white but "being set under starlight they would blaze with blue and white fires brighter than Helluin."

British author Doris Lessing (winner, at the age of almost 88, of the Nobel Prize for literature) produced a series of five books called *Canopus in Argos: Archives*, whose third volume is titled *The Sirian Experiments*. Published in 1980, *The Sirian Experiments* tells how agents of the empire of Sirius are allowed to try "evolutionary experiments" with humans in Earth's Southern Hemisphere. They are allowed by the agents of the more powerful empire of Canopus (the second brightest star), who conduct such experiments in Earth's Northern Hemisphere. The Canopeans call Earth "Shiskasta," a name which is said to imply abandonment and catastrophe, while what a literary critic of the work calls the "less perceptive" Sirians take a more hopeful (or naive?) view and call Earth "Rohanda, the Fair."

Even early twenty-first-century culture is still tapping into the fame of Sirius. One of the first two popular satellite radio stations is called Sirius. More interestingly and richly, the tremendously popular Harry Potter books include an important character named Sirius Black. Anyone who knows a little about the lore of the star Sirius has a pretty good clue, when the character is first introduced, what his secret nature will turn out to be.

The View from Sirius

As fictional as it may seem, some idea of what we would really see if we visited the Sirius system has been learned by science. If we were on or near Sirius itself, we would see the companion as a mere point of light that was equal in brightness to (but far more concentrated and intense than) Earth's Full Moon when Sirius B was farthest (several magnitudes brighter when closest). From the Pup, located as far from Sirius as Saturn or Uranus (depending on what part of its elliptical orbit it is on), Sirius would shine somewhat less bright than the Sun does from Earth—but still magnificently and blindingly bright. The Sirius system might not host any planets, due to disruptions of them that may occur because of the position of Sirius B (if Sirius does have some planets, they would have had only a few hundred million years to have developed and so would be unlikely to yet host life, even less likely to host intelligent life as of yet).

What would the stars look like from Sirius? There are a total of ten known star systems within 10 light-years of Sirius and thirty-five within 15 light-years. The closest to Sirius is also the most luminous within 15 light-years of it—Procyon. At a distance of only 5.2 light-years from Sirius, Procyon would glow brighter in the Sirian sky than Sirius does in our own. Our own Sun would shine a little less bright than 1st magnitude as seen from Sirius and appear in Hercules not too far from the considerably dimmed Altair. Somewhat fainter than the Sun from Sirius would be Alpha Centauri (9.5 light-years from the Dog Star). The only other star within 10 light-years of Sirius that would be bright enough to see with the naked eye would be Epsilon Eridani (7.8 light-years from Sirius).

The Once and Future Sirius

Sirius is near the behind-us point in the heavens, the Solar Antapex or Sun's Quit. We look out the front window of the spaceship of the solar system and see Vega; we look out the back window and see Sirius. Despite the fact that Sirius is behind us, it is for a while approaching us—currently at a rate of 8 km/sec (its space velocity—speed through three-dimensional space relative to us—is 19 km/sec).

How bright does Sirius appear from Earth over the vast period of time during which it is passing us? Sirius has shined brighter than magnitude 4.0 in Earth's sky for less than 2 million years and will do so for less than 2 million more. It is a 1st-magnitude star (magnitude 1.5 or brighter) from our planet for a total of about 750,000 years. Sirius has been Earth's brightest star (other than the Sun) for the past 90,000 years (when it took the title from Canopus) and will be for another 210,000 years (when it loses the title to Vega). Sirius will come closest to us 60,000 years from now, when it will be a little closer (7.8 light-years) than it is now and shine a little brighter (magnitude −1.64).

The illustration on page 103 shows the path of Sirius in the sky long-term. The period displayed is from 2 million years in the past (when it was a dim naked-eye star in what is now the constellation Lynx) to 200,000 years in the future (when it will still be very bright and already past the current position of Achernar). Notice that Sirius passed near the present sky location of Pollux a few hundred thousand years ago and very near the present location of its fellow Dog Star, Procyon, almost exactly 100,000 years ago.

Motion of Sirius

Sirius now

+50,000

+100,000

+200,000 +300,000 years

−50,000

−200,000

−500,000

−1,000,000 years

The path of Sirius through the heavens over a long period of time (note that the background stars would move, too—though few as far through the sky as our near neighbor Sirius).

Peerless and Dearest Sirius

If these thousands of words about Sirius are not enough for you, you can read more about it, the star of stars, in my book *The 50 Best Sights in Astronomy*. Of all the astronomical sights that are frequently visible to the naked eye, Sirius is surely one of the very greatest. And in that other book I describe the three most memorable and emblematic observations of Sirius I've personally made.

One of those observations was of Sirius at its culmination, highest in the south, at midnight. In our period of history, this midnight culmination occurs, amazingly, on the night of December 31 to January 1. So whereas the ancient Egyptians deliberately began their New Year at the time of Sirius's heliacal rising (for the first time, probably around 2782 BC), today we enjoy the coincidence of the king of stars being highest at the opening minute of each year.

The peerless and dearest star, Sirius, is first in the year just as it is first in the star lover's sight and heart.

9 CANOPUS

Canopus is the most radiant star in the brilliant southern constellation Carina. At an apparent magnitude of –0.62, it is the second brightest of night's stars, shining only about half as bright as Sirius but quite noticeably brighter than Alpha Centauri. And yet of all the twenty-one stars of 1st magnitude, Canopus is probably another second—second least famous.

Part of the reason for Canopus's relative lack of fame is that it's one of the "South Circumpolar Six," never visible as far north as 40° N, where so many of the world's people live. But why, among even the South Circumpolar Six, is only Achernar less famous than Canopus? The Alpha Centauri system is our closest stellar neighbor in space, and it forms a close pair in the sky with Beta Centauri, just as Alpha Crucis and Beta Crucis do together in the compact and dramatic Southern Cross. Canopus, by contrast, is a rather lonely star in the sky.

But just because Canopus is not known by most of the public doesn't mean that it is a star of limited interest. Nothing could be farther from the

Canopus

ARGO NAVIS

Argo and Canopus.

truth. The name of Canopus has been in use longer than almost any other star's. As a sun, Canopus is a rare type of supergiant—and easily the closest supergiant star to us. It is therefore the star of greatest apparent brightness in our sky off and on—mostly on—during an immense period of more than 5 million years. Furthermore, Canopus is a navigational star not only in the conventional way (one of several dozen traditionally used for this purpose by ship sailors on our planet) but in another, very modern way for which it is uniquely fit.

Second to Sirius

Chicago in the United States and Lyons in France are called "the Second City." In like fashion, we can call Canopus "the Second Star"—the second brightest in all the heavens, but somewhat overlooked in favor of the First

Star, Sirius. Canopus comes to its midnight culmination on December 27; the date of its 9:00 P.M. culmination is February 11. Both are only slightly before those of Sirius. But in other ways, Canopus is not the leader of Sirius.

Not only is Canopus about 0.82 magnitudes dimmer than Sirius, it has the misfortune of lying moderately close to Sirius—Canopus is closer to Sirius in the sky than to any other 1st-magnitude star. It lies about 36° farther south than Sirius and only 21' of R.A. ahead of Sirius. When both Sirius and Canopus are highest as viewed from around 35° S (Sydney, Buenos Aires, and Cape Town), Sirius is about 18° north of the zenith and Canopus about 18° south of the zenith. It must be a beautiful sight to have the heavens' two brightest stars somewhat more than a Big Dipper–length apart on either side of the zenith. But having them displayed together emphasizes the superiority of Sirius's brightness to that of Canopus.

How Far North Is Canopus Visible?

At least viewers in Australia, New Zealand, South Africa, Argentina, and Chile all get a great high view of Canopus.

This raises the question of how far north on Earth Canopus can be seen in our time. There's a theoretical answer but a different one for real life that is interesting for observers to try to determine. Canopus is ever so slightly farther north than −53° declination so, theoretically, it should culminate right on the due south horizon for a viewer at 37° N latitude. Due to atmospheric refraction's bending images a little higher and the possibility of observing from a mountain, Canopus might even be spotted a little farther north. On the other hand, atmospheric extinction (scattering and absorption of incoming light) is very great down near the horizon. On a very clear night, a star at 2° high is dimmed by 2.5 magnitudes and at 1° high by 3.0 magnitudes. In practice, clouds (even thin ones), haze, and also light pollution, would usually make the situation much worse. But it ought to be possible to see Canopus on some occasions when it is only 1° above the horizon, even with the naked eye. Use binoculars or a telescope, and your chance for success is even better.

Here in the eastern United States, I know of observers in South Carolina who see Canopus, but the star should also be visible from the rather tall mountains of North Carolina (whose north border is just above $36^1/2°$ N). I'd be glad to hear from readers of this book who see Canopus from near its limiting north latitude—or any bright star within a degree of the horizon.

What is the view of Canopus like from the southernmost parts of the continental United States—southern Florida and southernmost Texas? Even there, Canopus gets no higher than about 10° to 12°. On very clear nights, atmospheric extinction still dims it by about a magnitude, down to about the brightness of Procyon.

The United Kingdom and continental Europe are considerably farther north, as a whole, than the United States (Great Britain is at the same latitude as Labrador, but its climate is much milder due to the influence of the Gulf Stream). In Europe, Sicily and southern Greece are just within the possible observation zone for Canopus. But viewers there have to know exactly where and when to look, have superb weather, and possibly use optical aid. Basically, it is accurate to say that Canopus is not really visible—certainly not to a casual observer—from Europe. Was the situation different a few thousand years ago, due to precession of Earth's axis? No, it wasn't. Then, as now, Canopus was not really visible from Europe—not visible from ancient Rome or Greece—but came into view at the latitude of northern Egypt. And that fact is the basis for a remarkable story about the origin of this star's name.

"Canopus" as Star, Pilot, and Port

The first two syllables of "Canopus" are not pronounced like those of "canopy" (as, for instance, in the phrase "the starry canopy of the heavens"). The accent in the star's name falls not on the first but the second syllable: *kan-OH-pus*, not *KAN-oh-pus*. But where the name comes from is what's really interesting: it may be the only 1st-magnitude star named after a person.

According to Greek legend, Canopus was the name of the pilot of the fleet of Menelaus when that red-haired king (and husband of Helen) tried to sail home from the sack of Troy. When the ships came to Egypt, the story goes, Canopus went ashore and was promptly bitten by a poisonous snake and died. In his honor, we are told, Menelaus gave his pilot's name to both the port he founded there and the bright star that rose while he was making the dedicatory speech.

Of course, it's quite possible that Canopus was not a real person. It may be that he was made up by later Greeks to provide a colorful explanation for a star name whose origin they didn't understand. Present-day star-name expert Paul Kunitzsch says that Canopus was an untranslated proper name that was introduced into Greek astronomy as late as "perhaps in the 2nd century B.C." R. H. Allen says the name "apparently" was first given to the star by Eratosthenes, who lived in the third century BC. Kunitzsch states that there "seem to be Egyptian influences in the name's development."

Allen mentions that the geographer Strabo, who lived in the first century BC, said that the name Canopus was "but of yesterday." But, claims

Allen, the star was known as Karbana in the writings of an Egyptian priestly poet in the time of Thothmes III—roughly 1,500 years before Strabo.

Allen claims that another early name for Canopus was, in China, Laou Jin, "the Old Man," "an object of worship down to at least 100 B.C." The early Hindus identified Canopus with Agastya, one of their Rishis (sages), who was the helmsman of the ship Argha (sounds like the Greek ship *Argo*, whose constellation Canopus shines in)—a ship used to survive a great flood. (Agastya was the son of Varuna, the goddess of the waters, and after the deluge performed a magical purification of the water—Agastya, say lore experts Gertude and James Jobes, "cleared the waters spontaneously like the heart of one virtuous.")

Let's return to the topic of the port of Canopus. If the Trojan War has any basis in real history, it may have occurred in the twelfth century BC. About a thousand years later, the port of Canopus was a center for "vice and dissipation"—it was located 12 miles northeast of the great ancient city of Alexandria. But eventually Canopus was also the location for the observations of the famous first-century AD astronomer and compiler Claudius Ptolemy (the old idea that the Sun and other planets orbited around Earth is often called "the Ptolemaic system" after him). The modern village on the ruins of the old port of Canopus was called by Western historians Al Bekur or Aboukir when it figured prominently in Lord Nelson's Battle of the Nile in 1798 and Napoleon's victory over the Turks the following year. The name of the port is now more correctly spelled Abu Qir.

By the way, precession of Earth's axis has not greatly affected the southerliness of Canopus in the past few thousand years, but 12,000 to 13,000 years ago, a half of the full precession cycle back, the south celestial pole was moderately near Canopus. This made it a well-offset but brightest South Star in the same millennia that Vega was a closer-to-a-pole and brilliant North Star. Thus the axis of Earth pointed to extremely brilliant stars in either direction as the most recent Ice Age ground toward a close.

Canopus Venerated by the Arabs

For many centuries, the Arab peoples have had a great fascination and respect for Canopus. One name that they gave to Canopus was Suhail. Several bright stars seen low in the south by the Arabs were called Suhail or Suheil and Wezen. "Wezen" means "weight," presumably a reference to the way in which a bright star near the horizon would seem to be hanging low with the heaviness of its own great load of light. "Suhail" is said by star lore expert Kunitzsch to be of unknown origin. But astronomy writer and classical scholar Guy Ottewell (and at least one much earlier authority) supposes it is related to the Arabic *sahl*, which means "smooth" and "plain"—an example of which is the Sahel semidesert across Africa, south of the Sahara. Ottewell says that *suheil* is the diminutive form of this word, used as a boy's name and the name for Canopus. The idea of giving a star like Canopus such a name would presumably be that it is seen down near the horizon, just over the "smooth plain" of the desert stretching away from you. To use Suheil/Suhail as a name for a boy would, one supposes, only make sense if the intention was to name him after the star. This is supported by R. H. Allen's statement that "Suhail was also a personal title in Arabia, and, Delitzsch says, the symbol of what is bright, glorious, and beautiful, and even now among the nomads is thus applied to a handsome person." Allen also wrote that "[a]mong Persians Suhail is a synonym of wisdom, seen in the well-known Al Anwar i Suhaili, the Light of Canopus."

An Arab legend tells of a youth named Suhail who may be Canopus, who married the maiden Al Jauzah (Orion) but murdered her and then had to flee, chased by Sirius. Another version of the tale says that Al Jauzah rejected Suhail and kicked him unceremoniously to the south. Both of these versions presumably arose as attempts to explain Canopus's southerly separateness—a seeming exile—from Sirius, Orion, Procyon, and the other bright constellations of winter.

Allen says that among the Arabs the star Canopus was even believed "to impart their much prized color to their precious stones, and immunity from disease." In the previous chapter, we talked much about how the heliacal rising of Sirius warned the ancient Egyptians of the annual flood of the Nile as summer began and therefore was used by them to mark their New Year. But Allen claimed that the heliacal rising of Canopus "even now [1899] used in computing their [the Arabs'] year, ripened their fruits, ended the hot term of the summer, and set the time for the weaning of their young camels." Allen said that the last of these practices was alluded to by the poet Thomas Moore in his *Evenings in Greece*:

A camel slept—young as if wean'd
When last the star Canopus rose.

Another early title of Canopus among the Arabs was Al Fahl, "the Camel Stallion."

Canopus must have become a star of interest for nineteenth-century English poets because this was the time of the European rediscovery of ancient Egypt. In his "Dream of Fair Women," Alfred, Lord Tennyson has Cleopatra herself speak of "lamps which outburn'd Canopus."

Percy Bysshe Shelley used Canopus for his own purposes in his brilliant and charming poem "The Witch of Atlas":

> Like a meadow which no scythe has shaven,
> Which rain could never bend or whirl-blast shake,
> With the Antarctic constellations paven,
> Canopus and his crew, lay the Austral lake.

The Rudder of Lost Argo

Today, Canopus is Alpha Carinae, located in the constellation Carina, the Keel. But this keel was until modern times just part of the largest of all constellations given to us by the ancient Greeks: Argo Navis, the ship *Argo*.

Of course, the *Argo* is the most famous ship of Greek myth, the vessel used by Jason—and his Argonauts—on the quest for the Golden Fleece. A few of the heroes on the *Argo* are honored by their own constellations: Hercules has his own summer constellation, and the twin brothers, Castor and Pollux, are represented by the winter constellation Gemini. But by the nineteenth century, astronomers started to find that the constellation Argo was inconveniently large for their purposes. As a result, they classified stars as belonging to separate parts of the great ship. When in 1930 the International Astronomical Union voted to recognize eighty-eight official constellations, the roster included the three major parts of the old Argo: Carina, the Keel; Puppis, the Poop or Deck; and Vela, the Sails. Nearby stars had alternately been called Malus, the Mast, or Pyxis, the Compass (of the ship *Argo*), but it was the latter, created by the astronomer Lacaille in the eighteenth century, which received IAU certification.

An odd thing about constellations is that they often portray only part of the imagined thing they represent—Taurus is half a bull; Pegasus, a horse with wings but lacking hind legs in the sky. As large as Argo was, it seems to have been pictured as only the rear half of the ship. Even stranger, the ship sails backward across the sky each night, led by the star usually pictured as being in its rudder—Canopus. The brightest sections of Argo are those parts of Puppis, Vela, and especially Carina that are crossed by the Milky Way band—Canopus is actually in a fairly star-poor part of the heavens that precedes the rest of Carina, ablaze with so many bright stars.

The Greek alphabet used to designate stars in Carina, Vela, and Puppis is that which ran through the original Argo. In other words, Canopus, which even well into the nineteenth century was widely known as Alpha Argus (the Alpha star of Argo), is now Alpha Carinae. To cite another example, today's

Gamma Velorum (in Vela) was once Gamma Argus. There is no Alpha or Beta star in Vela or Puppis because the original Alpha and Beta stars of Argo are in Carina.

Canopus, the brightest star in Carina, once had the honor of being the brightest star in the largest of all constellations, Argo. Amazingly, however, there was a time—maybe a number of months—during which this second-brightest star in all the heavens lost even the title of brightest star in Carina. Canopus was in 1843 apparently slightly outshined by the most spectacularly bright of all variable stars that are not novae or supernovae: Eta Carinae. In April of that year, Sir John Herschel estimated the brightness of Eta Carinae to have reached –0.8. From the time of its discovery by Halley in 1677 until this 1843 outburst, Eta Carinae had been a strange *naked-eye* variable star, but in 1868 it faded out of naked-eye visibility and has typically been an 8th- or 7th-magnitude object ever since. But Eta could flare up again, and it is associated with one of the most spectacular of all deep-sky objects, the continuingly great Eta Carinae Nebula.

The Struggle to Find Canopus's Distance and Luminosity

Until rather recently, Canopus was often listed as shining at magnitude –0.72. It is now measured as –0.62. This, however, has not been the source of the tremendous amount of trouble astronomers have had in figuring out how luminous a star Canopus really is. The problem has been in establishing the distance of the star.

Canopus has the distinction of being the bright star whose true distance and therefore luminosity has been most variously and widely misjudged. As recently as the 1960s, authorities placed this star at over 600 light-years away, with a luminosity of 60,000 Suns. Burnham, in his *Celestial Handbook* (first widely published in the late 1970s), gave a correction to just 100 to 120 light-years and an absolute magnitude of –3.1—bright as about 1,400 Suns. He suggested a diameter about thirty times that of our home star. The early 1990s book *Starlist 2000.0* relied on then recent information from the Royal Astronomical Society of Canada's *Observer's Handbook*, which stated that Canopus was only 71 light-years away, with an absolute magnitude of –2.4, a luminosity of about 718× Sun and a diameter of about 13x Sun! Must Canopus therefore be in luminosity class II? Not according to *Sky Catalogue 2000.0*, whose 1991 edition had Canopus as a F0Ib supergiant that is 205 light-years away.

Canopus has indeed turned out to be a supergiant—a rare yellow-white one, a variety too poorly understood to have had its luminosity confidently estimated by spectral means and used to help calculate its probable distance. Only precise measurement of the parallax of Canopus could produce an accurate figure for its distance and answer key questions about its nature. Finally, in the 1990s, Hipparcos determinations of position enabled astronomers to at last get a measurable parallax for Canopus and therefore a distance.

Canopus as a Sun

Canopus is now estimated to lie 313 light-years from Earth. This puts its absolute magnitude at –5.4 and its luminosity at 15,000 times that of the Sun. Canopus is thus a luminosity class Ib supergiant—and the closest supergiant of any kind to Earth.

Canopus is a rare F-type supergiant—or, according to some experts, just out of the F type and an A9 star (that's considerably cooler than the also greatly more luminous A2 supergiant Deneb). Canopus must in any case be one of those supergiants that is now rapidly fluctuating across a range of spectral types. It could already have been a red supergiant—or may yet become one.

But will Canopus become a supernova? What will be its fate?

Jocelyn Tomkin, in his 1998 article on past and future brightest stars in our sky, suggested that Canopus just might go supernova before it lived to be Earth's brightest star again, between 480,000 and 990,000 years from now. But some more recent writings speculate that Canopus may fall just short of being massive enough to go supernova. Its luminosity suggests that it had a birth mass of about eight or nine times that of the Sun. So its fate may be like that of Sirius B—to lose most of its material and become a more massive than average white dwarf. Most white dwarfs are made of carbon and oxygen. But the massiveness of Canopus may enable it to fuse further and end life as a rare neon-oxygen white dwarf.

If Canopus's precise spectral class is either F0 or A9, its surface temperature should be about 7,800 K. What is its color? Procyon is reputed by some observers to have a tinge of yellow in its whiteness, as is Altair. The claim is made more often for Canopus. But in the B-V system of color measurement where 0 is blue-white and 2 is quite red, color indices are as follows: Procyon (+0.43), Altair (+0.22), Canopus (+0.16). So Canopus ought to be the least yellow of the three. Judging from what I've seen of Procyon and Altair, and what I know other observers have seen of their hues, I would say it's not impossible that Canopus could show the slightest hint of yellow to an observer with particularly sensitive color vision. But surely almost every one

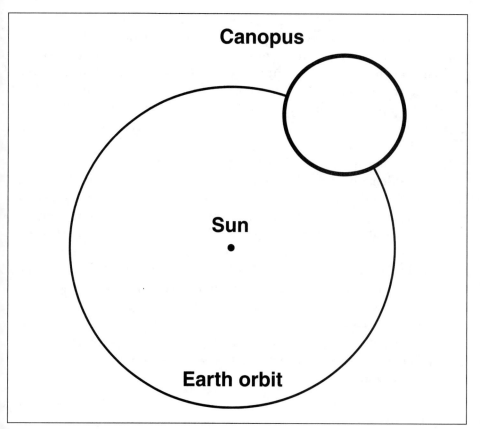

Canopus

Sun
•

Earth orbit

Relative sizes of
Canopus and
Earth's orbit.

of the many claims of seeing Canopus as yellow or slightly orange can be credited to its having been observed when reddened by its lowness in the sky.

How big is Canopus? Given its luminosity and mass, it should be about 65 times as wide as the Sun. Canopus much exceeds the Sun in another way. The Sun's corona (outer atmosphere) has a temperature of 2 million K. Canopus's corona must be about ten times hotter. It produces observable X-rays and radio waves.

Canopus the Four-Time King

Tomkin's aforementioned article lists sixteen different stars that either have been or will be the brightest in the Earth's skies between 5 million years ago and 5 million years hence. But only one of these stars is our brightest more than once—Canopus, which is king no less than four times.

Canopus first was brightest in our skies between 3,700,000 and 1,370,000 years ago. It passed the closest to us it ever will in that period: 3,110,000 years ago, it was only 177 light-years from Earth and burned at magnitude –1.86.

Then for a few hundred thousand years, two very much less luminous stars seized the title of brightest star—because they passed remarkably close (8 and 5.3 light-years) to us. But Canopus regained the throne 950,000 years ago, when it had receded to 252 light-years from Earth and shined at magnitude –1.09.

Canopus then held the title of brightest star until 420,000 years ago, when first Aldebaran and then Capella claimed the throne. But once these lesser suns had moved a bit farther away from us, Canopus reigned again. Starting 160,000 years ago, when it was 302 light-years away and magnitude –0.70, until 90,000 years ago, Canopus was brightest. Then, the reign of Sirius began. That reign will last until 210,000 years from now, when the Dog Star is exceeded by Vega. But when Vega loses the title, 480,000 years in our future? Then it will be Canopus—346 light-years away and magnitude –0.40—which will again be brightest star.

Not until 990,000 years in our future will Canopus be replaced as brightest star—and never again regain the title. But even 2 million years from now, Canopus will shine around magnitude 0. Even 5 million years hence, it will still glow at magnitude 1!

The Star to Steer By

A final fascinating distinction of Canopus has been its use, along with the Sun and bright planets, in interplanetary navigation. When I first heard there was a Sun Tracker camera and a Canopus Star Tracker camera on the various great unmanned American spacecraft of interplanetary exploration, I immediately guessed (correctly) why Canopus, as opposed to any other bright star, was being used as a navigational guide. The reason was something I had learned about while playing around with the positions of the brightest stars: as an adolescent studying my old edition of *Norton's Star Atlas*, I had discovered that Canopus is the 1st-magnitude star farthest from the ecliptic.

An interplanetary spacecraft usually finds the Sun, Earth, and other planets confined to nearly the same zone of sky as one another—from a side view the Sun and all the major planets stay within almost the same plane. And that zone is not far off from what on Earth is called the zodiac, centered on the ecliptic (the Sun's apparent path through the heavens). But how is the spacecraft to get its bearing on a direction perpendicular to, or nearly perpendicular to, that of the Sun? By keeping track of the position of the bright star located farthest from the ecliptic—Canopus.

Thus Canopus, a guide to wanderers in the Arabian desert, the rudder of the legendary ship *Argo*, and one of the navigational stars for sailors on Earth's seas, is another kind of navigational guide. It is a guide for the interplanetary sailors we call unmanned space probes—and maybe someday for human interplanetary sailors heading for Mars and other worlds of our solar system.

ALPHA CENTAURI

10

Alpha Centauri is the third brightest star in the night sky. But this far-south star is much better known for being the star that is closest to our own—a little more than 4 light-years distant. We really should say closest "star system," however, because a telescope easily reveals two bright stars that make up the single naked-eye point of light. And that's not the end of the inventory of stars in this system. If an observer knows where to look, a telescope also shows a dim third star a huge 2° away from the bright pair, and that star is also part of the system—in fact, it is the member of the system closest of all to us.

Fortunately, travel from the Northern Hemisphere to the Southern Hemisphere is now much more convenient and common than it used to be, and twentieth- and twenty-first-century science has taught us much more about Alpha Centauri. So it's time to start publicizing not only the nearness and brightness of Alpha Centauri but also the other remarkable facts about this star. Point a telescope—almost any telescope—at Alpha Centauri and you split it into by far the brightest of double-star duos. Likewise, Alpha forms—with Beta Centauri—the brightest of all compact star-pairings visible to the naked eye. And what science has learned is as interesting as those sights are striking: Alpha Centauri A is one of the stars most similar to our Sun, perhaps even to the point of possessing planets and, just possibly, life.

α Cen A & B

Proxima

Images of Alpha Centauri and Proxima Centauri. Circles in upper right close-up are meant to suggest the diffraction disks of Alpha Centauri A and B as seen through a telescope—not their true globes, which would be stupendously smaller on this scale.

The Closest Neighbor

We still should start our discussion of Alpha Centauri with a look at its most famous attribute: its nearness.

The story of how that nearness was discovered begins in the 1830s. Three different astronomers were trying to be the first to find stellar parallax, each astronomer with a different star. Friedrich Wilhelm Bessel settled on a dim naked-eye double with large proper motion, called 61 Cygni. Friedrich Wilhelm von Struve chose Vega. But Scottish astronomer Thomas Henderson had the advantage of being the director of the observatory at the Cape of Good Hope in South Africa—a place where he could observe Alpha Centauri.

Bessel completed his work before the other astronomers and gets the credit for being the first person ever to make a (reasonably) accurate measurement of the distance to a star. But Henderson finished second, and to him goes the honor of discovering the distance to what we still believe is the closest star system to our own—Alpha Centauri.

The story of Alpha Centauri's closeness doesn't end there, however. It wasn't until 1915 that proper motion studies of stars by R. T. Innes turned up an amazing discovery: the third, very dim member of the Alpha Centauri system, surprisingly far from the A and B pair. Because this third star had a larger parallax and proper motion than Alpha Centauri A and B, it was judged to be even a little closer to us than they are. This star, "Alpha Centauri C," gained a more popular name: Proxima Centauri, "the near (one) of Centaurus."

Does Proxima Centauri orbit Alpha Centauri A and B? The answer was only "probably" until a few years ago. But in 2006, confirmation was achieved. Proxima is now on the near side to us of what must be a roughly half-million-year-long orbit.

How near are these nearest stars to Earth? Immensely farther than the outermost planets of our solar system, but almost exactly half the distance of the second-closest star visible to the naked eye, Sirius. Alpha Centauri A and B are now known to be 4.40 light-years from us. Proxima Centauri is 4.22 light-years from us. Proxima is so far from A and B—about 10,000 AU, 1 trillion miles, or one-sixth of a light-year—that its distance from Earth is only about twenty times farther than its distance from A and B.

The Most Spectacular Bright Double

If you are visiting the Southern Hemisphere and see Alpha Centauri, you don't just have to ponder its closeness to get excited. With some optical aid, you can also reveal it to be the most spectacular bright double star in all the heavens.

The two bright members of the Alpha Centauri system together form a single naked-eye point of light that shines at magnitude –0.28. That makes it the third brightest star in Earth's night sky. But the two components, A and B, shine respectively at magnitudes of –0.01 and +1.35. So if they are considered separately, Alpha Centauri A is the fourth brightest nighttime star and Alpha Centauri B is the twenty-first brightest. Put them both on the list of brightest stars (something that is not usually done) and Alpha Centauri A comes in just behind Arcturus (magnitude –0.05) but ahead of Vega (0.03). And Alpha Centauri B comes in just ahead of Regulus (1.36), the dimmest of the 1st-magnitude stars.

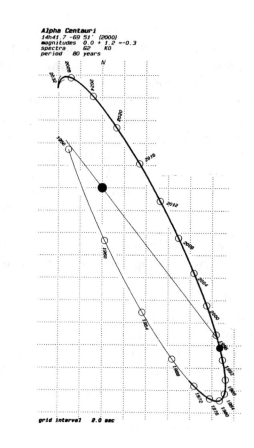

Orbit of Alpha Centauri B around A. The filled-in circle represents B's position in 1994.

Not only are Alpha Centauri A and B bright, they are usually separated enough to split with a very small telescope. At their greatest separation, the two lie 22" apart. But at the least separation during their eighty-year orbit around each other, the two do pull much closer together—to as little as 2" apart. The figure above shows that we see the orbit tilted about 11° from edge on and that the two suns will have the less extreme of their close pairings around 2015 to 2016. Not until more than two decades later will one of the closest encounters occur. At the start of 2006, A and B were 9.81", and at the start of 2007 they were 9.15" apart. They will be about 7½" apart at the start of 2009, 5½" at the start of 2012, and 4" at the start of 2015.

There is a claim that the duplicity of Alpha Centauri was first noticed by Jesuit missionaries in Africa in 1685. But the discovery is usually attributed to Father Richaud, who observed the doubleness of the star at Pondicherry, India, in 1689, during his observations of the comet of that year. (By the way, that comet had a strongly curved tail that extended as long as 68° on

December 14, 1689.) The first accurate measurements of Alpha Centuari A and B were made by Lacaille at the Cape of Good Hope in 1752.

How easy is it to observe the faint third member of the system, Proxima Centauri? Proxima shines at only magnitude 11.01 and lies 2.2° southwest of the A-B pair, so a medium-size telescope and a detailed finder chart are needed.

The Southern Pointers of Centaurus

Alpha Centauri is a glorious double star as seen in telescopes (with another companion far away). But it is *part* of a naked-eye pairing with the second-brightest star of Centaurus, Beta Centauri. This is a pairing in the sky, not space, because Beta Centauri is many times farther away than Alpha Centauri. Nevertheless, the two form a striking apparent duo in our sky. They are sometimes known as the Southern Pointers, because a line drawn from Alpha Centauri through Beta Centauri (almost due west) points toward the Southern Cross.

The separation between Alpha and Beta Centauri at the start of the twenty-first century is 4° 25'. How does it compare with the sky's other close pairing of 1st-magnitude stars (Alpha Crucis and Beta Crucis) and Gemini's pairing of a 1st-magnitude star and very bright 2nd-magnitude star (Pollux and Castor)? The table below lists separation, magnitudes, and difference in magnitudes of the stars within each pairing.

Pairing	Separation	Magnitudes	Difference in Magnitudes
Alpha Cru, Beta Cru	4° 15'	0.77, 1.25	0.48
Alpha Cen, Beta Cen	4° 25'	−0.28, 0.58	0.86
Beta Gem, Alpha Gem	4° 30'	1.16, 1.58	0.42

It's fascinating that the three tightest of the sky's pairings of very bright stars are almost equally tight. The Alpha Centauri–Beta Centauri pair ranks second in smallness of separation. What's really amazing, however, is that the large proper motion of Alpha Centauri will change the rank of these pairs not in thousands or in hundreds of years but in little more than 1½ centuries. Alpha-Centauri has the largest proper motion of any bright star—3.68" a year—and the motion is toward position angle 281°. Thus it is heading almost due west, almost straight toward Beta Centauri, which itself is much more distant in space and therefore has only a slight proper motion. By the middle of the twenty-second century, the apparent gap between

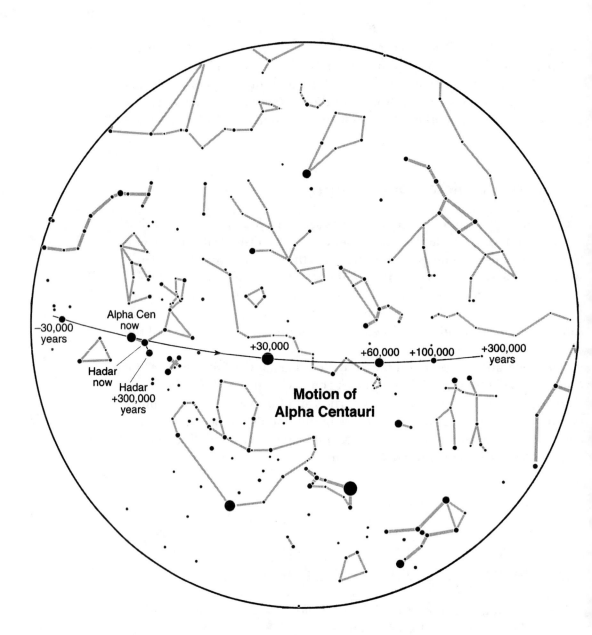

The path of Alpha Centauri through the heavens over a long period of time (note that the background stars would move, too—though not as far through the sky as our near neighbor Alpha Centauri).

Alpha and Beta Centauri will have shrunk to less than that between Alpha and Beta Crucis. Much farther in the future, Alpha Centauri will pass quite close to Beta Centauri in our sky, forming a naked-eye optical double with it for a while.

Let's step back and picture the scene when the Southern Cross and the Southern Pointers are near their highest.

First, these stars when highest are in the south—not quite overhead or in the north—whether you observe from Florida or New Zealand. They are located at around –60° declination. New Zealand extends no farther south than 47° S, the southern tip of South America to 55°.

The second thing to see, after noting the direction they are in, is that the Southern Pointers are only about 10° (one fist-width at arm's length) east of the Southern Cross. The Cross itself is almost exactly 6° tall. And, speaking of tall, this cross is aligned north–south and so is indeed upright when it is on the north–south meridian of the sky and highest ("culmination"). Regardless of how near due south in the sky the Southern Cross is, one in any case notices that the line of Alpha and Beta Centauri is just about perpendicular to the Cross's orientation.

So we have two stars brighter than magnitude 0.6 less than $4^1/2°$ apart pointing 10° away to a cross 6° high with two stars brighter than 1.3 less than $4^1/2°$ apart on the Cross's near side (Alpha Crucis is the south end of Crux the Cross, Beta Crucis is the star marking the east end of the transverse bar). Now imagine yourself in a reasonably dark sky and add to this basic scene of geometrically aligned gems the breathtaking background of a brilliant section of Milky Way spangled with scores of naked-eye stars, speckled with star clusters, structured by dark nebulae—especially the most famous of them all, the Coalsack, which lies just east of Alpha and Beta Crucis, between them and Alpha and Beta Centauri. Furthermore, if you look to the side of the Southern Cross opposite from Alpha and Beta Centauri, you find even closer to the Cross than those Southern Pointers the massive marvel of the Eta Carinae Nebula.

Everyone has his or her own favorites, but it's difficult to argue that there is a more splendid short section of the heavens than this one.

Little Bit of Lore

As I scan through book after book on starlore written by Northern Hemisphere authors, I find few legends concerning Alpha Centauri and Beta Centauri.

In his discussion of Alpha Centauri, R. H. Allen cites only a dubious claim about Alpha's heliacal rising and Egyptian temples, the usual inclusion of

Alpha among the various bright southern stars that the Arabs called Hadar, Wazn (or Wezn), and Al Muhlifain, and one more concrete and interesting point. This point is that Alpha Centauri was important in South China as the determinant star of the "stellar division" called Nan Mun, "the Southern Gate."

Allen's treatment of Beta Centauri does, however, include mention of a few tales involving Alpha and Beta Centauri together. The Bushmen of southern Africa knew the pair as "Two men that once were Lions." The aboriginal people of Australia knew them as "Two brothers who speared Tchingal to death." But Allen doesn't tell us who Tchingal was—only that the eastern stars of the Southern Cross (meaning just Alpha Crucis and Beta Crucis?) were the spear points that pierced his body. Could Tchingal be the Coalsack, with Alpha and Beta Crucis being the shining spear points emerging after having passed through his dark body?

Gertrude and James Jobes claim that desert nomads said Alpha and Beta Centauri were Khaiyal ("cavalier") and Zammal ("mule driver")—"two riders that come face to face across the lonely fields of the southern sky."

Alpha Centauri is the name—really just the Greek-letter designation—by which this third-brightest star in the heavens is commonly known. That's true, but one does fairly often see another name for it, not much livelier than the Greek-letter designation: Rigil Kentaurus. It is sometimes shortened to Rigilkent. It simply means "foot of the centaur." Interestingly, the four 1st-magnitude stars of Centaurus and Crux were all placed in feet or legs of the Centaur by Ptolemy in Egypt in the first-century AD. Back then, the Southern Cross was not recognized as a distinct constellation.

Yet another proper name applied to Alpha Centauri in modern times is Toliman, which is derived from an old Arabic term meaning "the ostriches." But no one knows which of the stars in Centaurus actually represented ostriches in early lore.

Far Centaurus and the Ambassador to Peladon

Of course, after it was recognized in the nineteenth century that Alpha Centauri was our closest neighbor, public interest in the star increased. If we look for cultural references to Alpha Centauri in twentieth- and twenty-first-century literature, movies, television, and video games, we can find many.

The first great science-fiction story in which Alpha Centauri played a major role may have been a 1944 tale by A. E. van Vogt. I read it in a much later anthology when I was a kid. The title of the tale—including the sound of that title—was what really filled me with admiration and has stuck with me ever since: "Far Centaurus." Although the name Proxima Centauri basically means "near Centaurus," the title of the story is appropriate because the tale tells of a first spaceship journey that would take many generations to complete—"'Tis for far Centaurus we sail!" I won't spoil the surprise ending of the story for you, in case you ever get a chance to read it. If you've just got to know, you can learn the ending on the Internet where, I find, I'm not the only one who has been stirred by that title. The title has caught on with at least one major spaceflight advocacy group. "Far Centaurus" is the title of an amateur astronomer's blog entry about his night of adventurous scanning for deep-sky objects down to near his south horizon in Centaurus.

You can find on the Internet various lists of numerous science-fiction references to Alpha Centauri. Strangely, the most extensive list I looked over missed the classic van Vogt story. It also missed the only important science-fiction references to Alpha Centauri on the very popular and long-running television series *Doctor Who*. In the "Monster of Peladon" and "The Curse of Peladon," the third Doctor takes his companion to the planet Peladon, where they encounter a gentle ambassador from Alpha Centauri. This being—described by one Internet commentator as a "hermaphroditic hexapod" (male/female walking on six feet)—was after its introduction to the protagonists regularly referred to as just "Alpha Centauri." The ambassador's appearance was hilariously preposterous. If I remember correctly, it was essentially a single huge pea-colored eyeball, which was situated in clothes something like a cupcake liner, had a high-pitched voice, moved with a dithering motion, and tended to worry a lot.

Alpha Centauri A and B as Suns

Among the hundred brightest stars (other than the Sun), there are only two star systems—Alpha Centauri and Capella—in which the primary star is of G spectral type like our Sun. But Capella A is a G6III star and huge, whereas Alpha Centauri A is a G2V star—just like the Sun. In fact, Alpha Centauri A is only a little bit larger, a tiny bit more massive, and a little bit more luminous than the Sun. Among the ninety-one stars other than the Sun that are closer than 19.15 light-years (those on a recent list I'm perusing, which happens to end at ninety-one), only two are G spectral–type stars—Alpha Centauri A and Tau Ceti. But Tau Ceti is a G8V star, considerably less luminous than the Sun.

What are the odds that among so many stars, bright in our sky or near in our neighborhood of space, the only one much like the Sun—in fact, very much like the Sun—would be our closest neighbor?

Again I'm amazed that the similarity of Alpha Centauri A to the Sun has not created far greater interest than it has. Perhaps one belief has worked against this fact's exciting scientists and laymen alike. It is the belief that a double- or multiple-star system might not be able to form planets or at least maintain ones in stable, roughly circular orbits (free from extreme temperature variations) where life could exist. Yet in recent years that belief has been overthrown—not only by new thinking about extrasolar planets but even the discovery of some in stable orbits in binary star systems.

Let's compare the Sun and Alpha Centauri A. The latter is believed to have a surface temperature of 5,770 K—only 10 K cooler than the Sun! Alpha Centauri A has an absolute magnitude of 4.34 compared to the Sun's 4.85, making it only about 60 percent more luminous than the Sun. Its mass is about 1.10 that of the Sun. Its diameter should be about 1.23 times that of the Sun. Alpha Centauri A's metallicity (content of elements heavier than hydrogen and helium) seems to be roughly twice that of our Sun. That and its luminosity suggest that it has been around considerably longer than the Sun, which was born about 4.8 billion years ago. Two different studies suggest that the Alpha Centauri system could have been born either 6.5 billion or 7 to 8 billion years ago. With its greater mass and age, Alpha Centauri A could be close to running out of the hydrogen fuel in its core, whereas the Sun can probably go at least a few more billion years before that happens. This situation shows that a small initial difference in stars—slightly greater mass—can lead to major differences further down the road.

But what of Alpha Centauri B? It is an orange K1V star, considerably cooler than its Alpha Centauri A or the Sun. Its absolute magnitude is only about 5.70, which would suggest that it is less than half as luminous as the Sun, but another source I consult says it has 51 percent of the Sun's luminosity. B is about 0.9 times as massive as the Sun and may have about 86 percent of the Sun's diameter.

We discussed the angular separation of Alpha Centauri A and B through telescopes. But what do those distances in seconds of arc work out to in actual space distance? Alpha Centauri A and B have an average separation of 24 AU but the distance varies—from as great as 36 AU to as little as 11 AU. Thus the gap between the two stars is, at average, similar to that between the Sun and Neptune. But it can dwindle to not much more than the Sun–Saturn separation.

Proxima Centauri as a Sun and Flare Star

Proxima Centauri is much different from Alpha Centauri A and B. A Hubble Space Telescope study found that Proxima is larger than once thought, but still only about 15 percent as wide as our Sun and about 20 percent as massive. And in visible light, this little red dwarf shines about twenty thousand times dimmer than our Sun—an absolute magnitude of 15.42. However, the surface temperature of this M5.5 star is only 3,100 K (its B-V color index is a very ruddy +1.94), so the whopping majority of its energy is emitted as infrared radiation. If we count this infrared glow, Proxima becomes about $\frac{1}{500}$ as luminous as the Sun.

What's most interesting about Proxima (other than the accident of its being the closest star to our own) is its status as a "flare star." The flares of Proxima Centauri may brighten it by as much as a magnitude in only a few minutes. The energy is similar to that produced by a flare on our own Sun, but in a star initially as faint as Proxima may represent a doubling in brightness. These flares are unpredictable but common enough for a few to occur in a year. No red dwarf has had as many of its flares observed as Proxima Centauri, but some red dwarfs produce mightier ones (for instance, UV Ceti has had flares that increased its brightness by as much as 5 magnitudes). In 1998, the Hubble Space Telescope apparently identified three cooler areas on Proxima, presumably "starspots," the equivalent of sunspots on our own Sun.

Is there evidence that Proxima Centauri is being orbited by an even tinier "brown dwarf" or by a planet? One study's claim of the former seems to have been disproved. But another study found a seventy-seven-day variation in the proper motion of Proxima that could be explained by a planet with 80 percent of Jupiter's mass orbiting only about 0.17 AU from the star. The researchers noted that there was a roughly 25 percent chance of a "false positive" in their data.

The View from Alpha Centauri

Suppose we found ourselves on a star orbiting somewhere around Alpha Centauri A or B. What would we see in the night sky? Amazingly, Proxima Centauri would shine at only magnitude 4.5. But at least its proper motion would be remarkable. The next closest star would be the Sun. It would appear as a magnitude +0.5 star not far in the sky from the Double Cluster in Perseus. In addition to the Sun, five other star systems are known to lie within 10 light-years of the Alpha Centauri system. The brightest of these, of course, would be Sirius. But Sirius lies almost a light-year farther from Alpha Centauri than it does from us, so it would be a little dimmer, a magnitude −1.2 jewel in the Alpha Centaurian sky. Most interesting of all would be the position of Sirius in that sky. It would shine about 2° west of Betelgeuse in an otherwise almost-unchanged Orion.

11 ⟡ ARCTURUS

My earliest memories of Arcturus don't go back as far as those I have of Sirius and Orion.

I was surely only seven or eight, however, when I stood out near the country road in front of my lifelong home in early spring and saw, poised at the top of a small oak tree just north of east, a brilliant and orange star. I knew from my books and a little star finder (planisphere) that this was Arcturus. For at least a few years, it became a tradition for me to greet Arcturus as it climbed from the point of this pinnacle. But the tree was dying, and one year, without warning, my father went out and cut it down, probably before its time. I didn't know his thinking in the matter so I couldn't then and can't now quite blame him. But decades later "the Arcturus Tree" stands there still in my memory each year as I greet the great glad star of spring.

Arcturus is a lot more than just the brightest star of the north celestial hemisphere, the brightest star of (Northern Hemisphere) spring, the second brightest star ever visible to most of the world's people. In one very important way, Arcturus is the most distinctive of all the 1st-magnitude stars. The strangest thing about Arcturus is where it originally may have come from. I'll save the surprise about this until the end of the chapter. Let's begin with another fascinating topic—the lore and legend associated with Arcturus.

Spring's Bright Bear-Guard

Legend begins in language (and language in legend). The legends about Arcturus are no exception. The name Arcturus is closely related to *arctic*—but not because Arcturus is anything like the most northerly of the 1st-magnitude stars (four of the other twenty stars are much farther north). Instead, because of legend . . .

The ancient Greek word *arktos* means "bear," and the star's ancient Greek name *Arktouros* meant "the Bear Watcher" or "Bear Guardian"—in reference to the star's mythic relationship to the neighboring but much more northerly constellation Ursa Major, the Great Bear. The fanciful idea that has come down to us is that Arcturus or its entire constellation Boötes, the Herdsman, is guarding a flock from the ferocious bear. In doing so, the herdsman has the

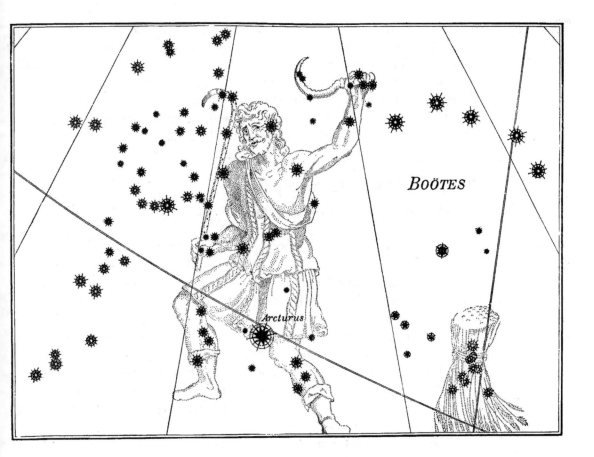

BOÖTES

Arcturus

Boötes and
Arcturus.

help of two starry hounds, imagined from ancient times but not turned into the constellation of Canes Venatici, the Hunting Dogs, until Hevelius did so in the seventeenth century. What is the flock that the Bear-Guard is guarding from Ursa Major? The flock has been said to be all the other constellations, which would otherwise be terrorized by the fierce Bear. (Incidentally, the star-name authority Paul Kunitzsch notes that the ancient name Arcturus was "reapplied" to the star in Renaissance times and that an alternative meaning of the name is "Guardian of the North." The latter is based on the interpretation that *arktos* means not just "bear" but "the land of the bear"—the north.)

Arcturus is mentioned by Hesiod and Homer and by numerous classical authors (Aratos, Hippocrates, Virgil, Horace, Plautus, and Pliny, to name a few). But is Arcturus mentioned in the Bible? If so, it would be the only individual star named in the entire work. The King James Version of Job 9:9 says that God is the one "which maketh Arcturus, Orion, and Pleiades, and the chambers of the south." The King James Version also has Job 38:32 as: "Canst thou bring forth Mazzaroth in his season? or canst thou guide Arcturus with his sons?" But later scholars seem to regard the references to Arcturus as a mistranslation for "the Bear." The Revised Standard Version has, "Can you lead forth the Mazzaroth in their season, or can you guide the Bear with its children?" The Bear is here intended by the translator to be Ursa Major—but one wonders if the Big Dipper stars were what was meant (or perhaps just the bowl of the Big Dipper, so that the stars of the handle could be the Bear's children?).

Bow, Owl, Algon, Hawaii-Finder, Fair-Opener, and Light to Hell

Arcturus and its constellation Boötes have, of course, inspired different legends and images in many cultures. One fanciful modern imagining of some people (myself included) is that Arcturus is a colored bow at the bottom of a kite (the body of Boötes is kite shaped). This kite is one that is flying high on spring winds.

Also from our own time is advice that beginning skywatchers are often given: "Take the arc to Arcturus and drive a spike to Spica." This means first to follow outward the curve of the Big Dipper's handle until you come (about one Big-Dipper-length onward) to Arcturus. A straighter line—a "spike"—from Arcturus leads on to the second-brightest star of the spring constellations, Spica.

A wonderful widespread legend of American Indians brings to life and notes in richer detail the Big Dipper–to–Arcturus curve. This legend tells that the bowl of the Big Dipper is a bear pursued from spring to fall by a trail of hunters. These hunters are the curve of stars that are today the handle of the Big Dipper and the west side of Boötes. The hunter stars are Robin (Alioth), Chickadee (Mizar), Moose Bird (Alkaid), Pigeon (Seginus in Boötes), Blue Jay (Izar), Owl (Arcturus), and little Sawwhet Owl (Muphrid). Not until autumn does the Bear weary, get wounded by the hunters' arrows, and stagger down to near the northern horizon—where his blood spatters to stain Robin's breast red and falls to encrimson the forests just below him (autumn foliage color change). But each year, the bear escapes (the hunters that were Boötes's stars all set), hibernates, heals all winter, and rouses in spring—only to have the pursuit begin again.

An even more delightful American Indian tale features Arcturus as Algon (White Hawk), who falls in love with the fairest of a dancing circle of maidens, the stars we see as Corona Borealis. (See pages 201–202 of my book *Wonders of the Sky* for a fairly detailed telling of this story.) South of the equator, Polynesians knew Arcturus as "the Star of Joy." They used it to navigate to Hawaii, for Arcturus passes through the zenith of that island chain. The journey back was achieved with the help of Sirius, which is the zenith star of Tahiti. (Although "40° N chauvinists" often celebrate Vega as the world's prime zenith star, Arcturus serves the purpose for Khartoum, Mecca, Bombay, Calcutta, Hanoi, Hong Kong, Taipei, Honolulu, Mexico City, Port-au-Prince, Kingston, and San Juan—even Miami and Brownsville, Texas.)

Legend and fancy have not ceased to become attached to Arcturus in modern times. A real-life event that has taken on the air of legend is the opening of the "Century of Progress" Exposition in Chicago in the spring of 1933. The light of Arcturus was focused by telescopes on photoelectric cells whose current activated a switch to turn on the floodlights at the exposition. Why was Arcturus chosen to open the event? A previous important Chicago fair had occurred in 1893, and astronomers in 1933 thought that light took forty years to get to us from Arcturus (we now know the correct distance to Arcturus is about 37 light-years).

Among various twentieth- and twenty-first-century science-fiction stories, UFO cults, and video games that mention Arcturus, perhaps the most notable is David Lindsay's 1920 novel *A Voyage to Arcturus*. This often heavy-handed and woodenly written work is also sometimes stunningly original—and it is ostensibly set on a planet circling Arcturus. The light of Arcturus in the novel is regarded as somber or ominous in keeping with the tone of the story, which I (but not all readers of the novel) interpret to be nothing less than the agonizing account of a soul's descent into hell.

Three Great Comets and "the Faire Starre Arcturus"

No bright star in modern times has had such wonderful conjunctions with mighty comets as Arcturus. In 1618, there were two notable comets. The initial comet had a bluish-white tail that reached a maximum length of at least 15°. It holds the distinction of being, as far as we know, the first comet ever observed in a telescope—a feat achieved on September 6 by none other than Johannes Kepler. But this was not the truly great comet of 1618. That was the later one, which was visible in broad daylight and then moved out to display a reddish hue in a dust tail that eventually extended to 104° in length! Both King James I and the diarist John Evelyn were so impressed they believed that this comet had presaged the Thirty Years' War. But it was when the comet's head passed roughly 5° from Arcturus that another person penned the most beautiful line about the comet. He was John Bainbridge, "Doctor of Physicke," and he wrote, "The 27th of November, in the morning, the comet's hair was spread over the faire starre Arcturus."

Speaking of beautiful: 240 autumns later, a comet that has often been called the most beautiful ever seen was near peak brightness and at a stage of lovely tail development just as it had an incredibly close encounter with Arcturus. This was 1858's Donati's Comet. It was the first comet ever photographed—on the night of September 27, when it was already approaching Arcturus. I wrote about the comet and its encounter with Arcturus in my book *Comet of the Century*.

This comet displayed marvelous hoods or envelopes in its head in telescopes and both this and the unusual gas tail have reminded people of 1970's incredibly beautiful (though less giant-tailed) Comet Bennett. My book *Comet of the Century* includes four marvelous old sketches of the entire Donati's Comet. The two most artistic and striking show the comet on October 4 and October 5—the first when it was over Notre Dame de Paris, the second—a lithograph possibly by Mary Evans—when it was over the Conciergerie in Paris. These dates were, respectively, the night of the approach of Donati's head to Arcturus and the night of its closest encounter with Arcturus, and both illustrations show well Arcturus and other stars in that part of the sky.

The final great comet to share a stage of glory with Arcturus came 137½ years after Donati's. It was 1996's Comet Hyakutake. On the American night of March 22–23—just a few nights before closest approach—Hyakutake zoomed only fairly near Arcturus, its close and therefore huge head instead passing almost right over Eta Boötis (Izar). Earlier that evening I saw in my 10-inch telescope a star of perhaps 6th magnitude get occulted by the comet's central condensation. But the naked-eye view of Hyakutake was greatly enhanced by the presence of Arcturus on that and the next few nights. Those next nights, some observers rated the comet's head as being just brighter than Arcturus. I was one of them, estimating a peak brightness of –0.4 for Hyakutake on two nights. Hyakutake offered several attractions Donati didn't (a huge, blue head; fast movement; and blue brightest part of tail, not to mention all-night, overhead-for- 40°-N, and Polaris-passing visibility). But even though its tail lengths could be said to compare favorably with Donati's (though only briefly), Hyakutake's tail was a rather narrow, relatively dim thing compared to the broad, complex, rich, and bright tail of Donati's—the comet whose head almost occulted Arcturus.

Champagne Shot with Roses

Arcturus is not only bright, it is colorful. Almost everyone agrees that even just to the naked eye, this star is not merely white. But there is no very bright star whose hue people have had as much trouble naming as Arcturus's.

If we look at an objective measure of color, the B-V color index, we find that among the 1st-magnitude stars, the +1.23 of Arcturus falls in a big gap between Pollux (B-V +1.00) and Aldebaran (B-V +1.54). Pollux is often said to be slightly orange and Aldebaran more deeply orange, though in reality even the supposedly "ruddy" Betelgeuse and Antares (both B-V +1.84) are, even to the more red sensitive of us, most fairly described as "golden-orange." Ptolemy wrote that Arcturus was *subrufa*—"slightly red." What do most observers call it? Burnham says "golden yellow" or "topaz" but mentions that Admiral Smyth called it "reddish yellow." One observer has compared it to the color of high-pressure sodium streetlights (that's painful to hear!). Whatever the hue, it is like that of no other very bright star and seems hard to pin down. My old friend Nora McGee had a memorable assessment, however. The color of Arcturus, she decided, is "champagne shot with roses." That would seem appropriate for a spring star that doesn't actually have its 10:00 P.M. (daylight saving time) culmination until June—the month of roses-and-champagne-celebrated weddings.

I believe that part of the confusion about the color of Arcturus is in trying to compare it with markedly dimmer "orange" stars such as Aldebaran and Pollux. I have a theory that the ideal naked-eye brightness for maximizing color perception is about magnitude 0.0 to –0.5—at least for my own color receptors. Color is washed out when the light source becomes too bright. This would explain why (independently of any Martian dust storms) I always see Mars as less colorful when it brightens beyond about –0.5 or –1; why Venus looks less yellow than Jupiter despite their essentially identical B-V values; and why Vega looks bluer than Sirius (though I and many other observers—dating at least back to Manilius—do see some blue in Sirius when the star is high enough on a steady night for our atmosphere to not ruffle its light into darting colors).

Arcturus as a Bright Star and a Sun

As recently as forty years ago, Arcturus was listed in books as the sixth brightest of all (nighttime) stars. Now we know that it is actually the fourth brightest, very slightly brighter than Vega and Capella. The apparent magnitude of Arcturus is –0.05—only a little dimmer than Alpha Centauri. However, the latter is a rather wide double star, and Arcturus slightly outshines the brighter of the two components. Interestingly, in the 1990s, data from the Hipparcos satellite seemed to prove something previously unsuspected: that Arcturus has a companion—supposedly about 0.25" from the primary and, if I understand the journal abstract correctly, a suspiciously bright

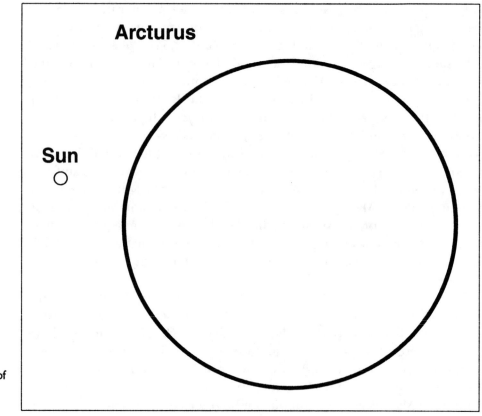

Arcturus

Sun

Relative sizes of
Arcturus and
the Sun.

magnitude 3 to 3½. Several different kinds of more recent investigations appear to have convincingly refuted the supposed existence of this companion. Arcturus seems to be a single star. But we have in recent years confirmed that Arcturus is a variable star—although varying by only 0.04 magnitude every 8.3 days.

Arcturus is reputed to have been the first star (other than the Sun) ever observed in a telescope in daytime—by Jean-Baptiste Morin in 1635. I have a friend who has observed Arcturus with the naked eye right at sunset. But in the nineteenth century, the great visual observer J. F. Julius Schmidt managed to see Arcturus with the naked eye 24 minutes before sunset.

At its distance of 36.7 light-years, Arcturus has an absolute magnitude of –0.3. Its luminosity in visible light is about 113 times that of our Sun. If you add in infrared light, however, Arcturus has luminosity 215 times greater than the Sun's. (Robert Burnham Jr. mentions that the amount of heat that sensitive instruments measure from Arcturus is equal to that from a single candle about 5 miles away.) As a giant star of spectral type K1.5 or K2, Arcturus has moved on from burning hydrogen to burning helium and has a surface temperature of about 4,290 K, considerably cooler than the Sun's. (Burnham, by the way, suggests that the spectrum of Arcturus "rather resembles the spectrum of a sunspot.") Calculations based on the luminosity and spectral type of Arcturus suggest the star should be about twenty-six times wider than our Sun. Direct angular measurement of the disk of Arcturus finds it 0.0210" across—which works out to its being twenty-five times wider than our Sun in space, about 20 million miles across.

Yet there are a few unusual properties of Arcturus. Its luminosity is actually somewhat greater than expected (could this be for the same reason as Vega—the brighter, hotter polar region of the star is pointed toward us?). A giant star with the luminosity and spectral type of Arcturus ought to have about four times the mass of the Sun. But recent estimates place the mass at no more than 1.5 times solar. Could Arcturus have already lost much of its mass in a strong stellar wind, as many red giants do? No, it seems to be early in its red giant phase. We can state that with the mass it currently seems to have, the ultimate fate of Arcturus should be to become a white dwarf, produce a planetary nebula, and ever so slowly fade away—much like our own Sun. What's more uncertain and far more interesting, however, is the star's earlier life and its origin.

Two unusual characteristics of Arcturus are very telling. First of these is a profound difference in the composition of Arcturus compared to the Sun and most stars we see in the sky: Arcturus is extremely metal-poor (meaning poor in elements heavier than helium), relative to the Sun. What this probably means is that Arcturus is a Population II star—a type of star born in a time or place unlike that of Population I stars like our Sun. Some Population II stars were born before the galaxy became much enriched in heavy elements from billions of years of supernova explosions. Population II stars are typically denizens of a spherical halo (centered around the hub of the galaxy but extending to great distances) rather than being residents of the flattened equatorial disk of the galaxy that contains its spiral arms—and Population I stars, including our Sun.

The other unusual characteristic of Arcturus is also strong grounds for thinking that Arcturus is a Population II star: its highly inclined orbit around the galaxy.

It Came from Beyond

The first clue of something surprising about the motion of Arcturus was uncovered by Halley. In 1718, he announced the discovery of "proper motion"—the displacement of stars against the background of very distant objects over the course of time. Halley compared the positions of Sirius, Arcturus, and Aldebaran from the sixteenth and seventeenth centuries to those recorded in ancient times and found that all three had changed. Arcturus had changed position the most. Modern studies show that the star's proper motion is 2.29" per year (in position angle 209°—that is, to the south-south-west). That means it moves well over 1° in 2,000 years. We now know that of all 1st-magnitude stars, the only one with a larger proper motion than Arcturus's is Alpha Centauri—but Alpha Centauri is almost nine times closer to us. By study of the Doppler shift in the spectral lines of Arcturus, we can determine that it is approaching us but that this component of its space motion is relatively slight—only about 5 km/sec. In contrast, the star's velocity through space relative to us is 122km/sec, much greater than that of any other 1st-magnitude star. Arcturus is passing us almost at a right angle, dropping through the Milky Way's equatorial disk in which our Sun and almost all the other stars we see in the sky are traveling. Exceptions from among countless thousands of stars are fifty-two others that seem to share the motion of Arcturus and are called the "Arcturus Group."

For decades, various books—including Burnham's—have been saying that Arcturus came close enough to be visible to the naked eye only 500,000 years ago and will have moved far enough away 500,000 from now to again fall below naked-eye visibility. Actually, the situation is not quite so drastic. A careful study using Hipparcos satellite data was executed by Jocelyn Tompkins (*Sky & Telescope*, April 1998, 59–63). His graphs show that 500,000 years ago and hence Arcturus was/will be still a bit brighter than magnitude 4.0. Of the five currently brightest stars, only one has a sharper spike up to and down from maximum brightness, and that is Alpha Centauri—a greatly less luminous star than Arcturus that is bright only while it is passing very close to us. How soon will Arcturus make its closest approach to us? On these vast timescales of hundreds of thousands of years, the surprising answer is only four thousand years from now: in no more future time than there has passed between early Egyptian or Sumerian civilization and the present, Arcturus will be at its closest to Earth. But, Tompkins calculates, Arcturus will then be only one-tenth of a light-year closer than it is now—and therefore imperceptibly brighter. Right now, we're seeing Arcturus at the best we ever will (in this 200-million-year circuit of ours around the galaxy, at least).

Many years ago, upon learning of Arcturus's highly inclined orbit around the galaxy and the way it is merely "dropping by" (literally) for a visit, it occurred to me to call Arcturus "the Gypsy Star." This was in tribute to its own wandering nature—not because I believed that gypsies paid special attention to it (although one could well imagine painted caravans moving

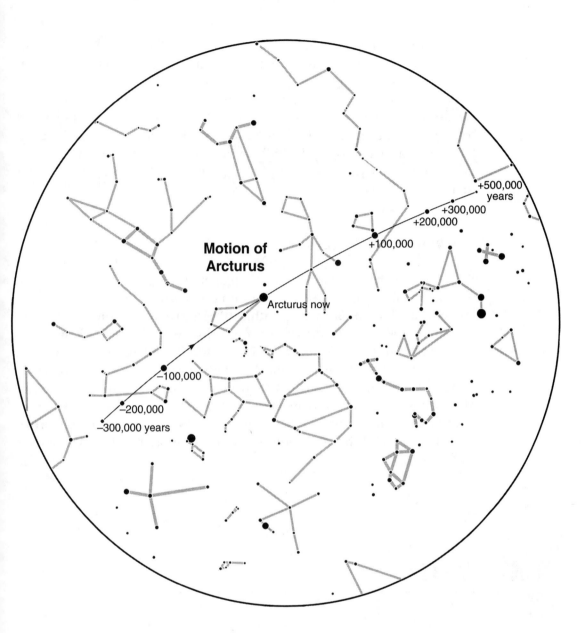

Motion of Arcturus

+500,000 years
+300,000
+200,000
+100,000

Arcturus now

−100,000
−200,000
−300,000 years

The path of Arcturus through the heavens over a long period of time (note that the background stars would move, too—though few as far through the sky as our near and truly fast-moving neighbor Arcturus).

out in spring beneath the light of Arcturus). But Arcturus may be a visitor from even farther away than we thought. It may not just be a Gypsy that has wandered here and there around our galaxy.

A new theory is that Arcturus and the other members of the Arcturus Group are members of a galaxy that collided and merged with the Milky Way sometime between 5 and 8 billion years ago—well before the birth of our Sun and solar system about 4.6 billion years ago. The fourth brightest star in our heavens today may actually be a refugee from an "island universe" other than our own!

The View from Arcturus

Robert Burnham Jr. noted that from a planet orbiting Arcturus, virtually all the stars would appear to be moving with great proper motion. There turn out to be ten known star systems within 10 light-years of Arcturus (similar to the number within 10 light-years of our own Sun). The closest star of any kind to Arcturus and the only bright star within 10 light-years of it is Eta Boötis—Muphrid. This magnitude 2.68 star is located just a few degrees west of Arcturus. Surprisingly, its distance from us is identical to that of Arcturus. Muphrid is only 3.3 light-years from Arcturus. For an Arcturian, Muphrid would appear about magnitude –2½ (similar to Jupiter seen from Earth). From the Muphrid system, Arcturus would appear brighter than Venus does from Earth.

 # 12 VEGA

Vega is easily the brightest star of the traditional summer constellations. It is the third-brightest nighttime star visible to most of the world's population. It also passes virtually overhead for those people who live approximately around 40° N latitude. Add finally that Vega may be the most noticeably bluish star for naked-eye observers and you can see why I like to call this star by two titles: "the Queen Star of Summer" and "the Sapphire of Summer." Vega's frosty hint of blue is refreshing to the mind and spirit after the end of a long, hot summer day. And early on summer evenings it shines down to us from the zenith, turning that spot into the exalted throne of the sky.

The Queen Star of Summer

In an article on the Summer Triangle I published in the now sadly departed *Night Sky* magazine, I wrote the following: "Vega should be utterly familiar to anyone who looks up on summer nights. This star is as much a part of my summer life as heat lightning and fireflies, fresh strawberries and corn on the cob, sunflowers and Queen Anne's lace—even as much a part as the sweat on my brow."

Vega is surely one of the two quintessential stars of (the North Hemisphere) summer, the other being Antares. In contrast to low and somewhat ruddy Antares, however, blue-white Vega is the brilliant star that passes near the zenith for more people in the world than any other. Winter's Capella is only marginally less bright than Vega and passes high in the sky at midnorthern latitudes, but attention is distracted from it in winter by so many other bright stars (including much brighter Sirius) and constellations (such as Orion). Vega is the fifth-brightest star in all the nighttime heavens, but third brightest properly visible to most of the world's population. At midnorthern latitudes, only Sirius outshines it greatly, but Sirius belongs to the opposite time of year. A friend of mine has, in fact, called Vega "the Sirius of Summer."

The second brightest star of northerners, Arcturus, is visible during many of the prime months of Vega, but it is only very slightly brighter than Vega. And it is Vega that reigns overhead on evenings in the summer, when more people are outside at night than in any other season.

Vega in the Summer Triangle

Vega has another advantage over Arctrurus and other stars: it dominates by brightness and position both its conspicuous (because compact) constellation and a conspicuous (because bright and big) asterism. Vega's small constellation is Lyra; the huge asterism is the Summer Triangle. The pattern of Lyra is a tiny (about 2° per side) equilateral triangle that sits on top of a slightly bigger north–south elongated parallelogram. Vega enormously outshines other stars of Lyra and shines on the northern end of this pattern— the "top" of it for most observers, at least during its long climb to the zenith in the summer months. Vega is also at the top of the giant transconstellational asterism called the Summer Triangle during the many hours of summer

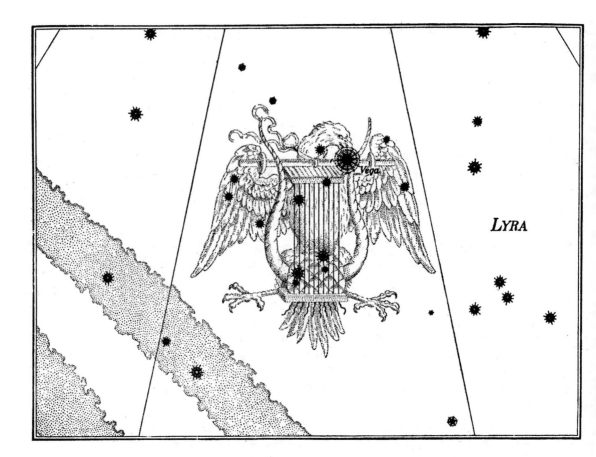

LYRA

Lyra and Vega.   evenings when the Triangle is rising. And Vega easily outshines the other two bright stars of the Summer Triangle, being about twice as bright as Altair and about thrice as bright as Deneb.

It is interesting to note that Vega and the Summer Triangle linger long past summer. Vega reaches its acme at midevening in late July and early August, when summer is at its culmination, at its fullest and richest in many ways. (At least where I live, I associate this midevening summit-reaching of Vega with local berries reaching their ripeness and meteors, too, being at their richest, their most plentiful and ready for the picking.)

But the Summer Triangle as a whole reaches its midevening culmination even later, at summer's very end. Then, as nightfall gets later and later, we find the Summer Triangle still prominent after dusk as it slowly descends in the western sky. At midnorthern latitudes, more southerly Altair gets lower faster than Vega and so "catches up" in the journey across the sky, coming to be at the same altitude as Vega and making the Vega–Altair line the horizontal base of the Triangle. Then Altair gets lower than Vega, leaving Vega and Deneb to shine prominently, though low in the west, at December nightfalls.

I vividly recall what time of year Vega rises in the northeast soon after evening twilight, for observers around 40° N latitude. In April 1997, when I was watching the mighty Comet Hale-Bopp get low in the northwest, I was delighted to see Vega come above the northeast horizon just in time to act as an excellent comparison star for the then-zero-magnitude comet.

Vega in the Telescope

There are famous telescopic attractions in Lyra—the Ring Nebula and Epsilon Lyrae (the Double Double)—but handbooks seldom have Vega mentioned as an object for telescopic observation. Yet it is one, and not just for the thrill of seeing one of the brightest stars made apparently many magnitudes brighter in the eyepiece. There is also the optical companion of Vega.

"Optical companion" means the star is not physically connected to Vega, is at a different distance, and just happens to be on nearly the same line of sight as the bright star. But the view is the same as if the companion were physical: a good 6- to 10-inch telescope on a summer night of steady seeing shows within Vega's throbbing wash of sapphire radiance an intense little spark of light just over 1 arc minute south of Vega. This companion shines at about magnitude 9.5—thus almost ten thousand times dimmer than Vega. (Incidentally, there is a star of similar brightness located about twice as far from Vega and to the northeast, not south, of Vega.) A real challenge, fit for larger amateur telescopes, is an 11th-magnitude star slightly less than 1' west of Vega. Like the magnitude 9.5 star to Vega's south, this object is slowly getting farther in the sky from Vega as the decades and centuries pass.

Vega as Destination, Pole Star, and Future Brightest Star

We have to deal with time periods much longer than just centuries if we want to discover some of Vega's most interesting distinctions.

First, there is the fact that our Sun and its solar system is heading toward Vega. Only roughly so, and how roughly depends on what sampling of local stars we use to determine our Sun's motion. But if we imagine our whole solar system as a spaceship orbiting around the center of our Milky Way galaxy, we can say this: Vega is the bright bluish star out the front window of our vehicle and Sirius is the bright bluish star out the rear window. This does not mean that our solar system will ever arrive at Vega—for Vega itself is going around the galaxy and we are merely heading for the place in space it now occupies. It will take our solar system a little less than half a million years to cover the 25 light-years that is now the gulf between us and Vega.

What really happens over hundreds of thousands of years is significant changes in the distance between us and Vega. Currently, Vega is getting closer to us. The Hipparcos satellite data from the 1990s now enable us to determine that Vega will eventually dethrone Sirius as the star of greatest apparent brightness in the skies of Earth. Vega should hold the title throughout the period from 210,000 to 480,000 years from now. It will come as near as 17.2 light-years around 290,000 years from now and then shine at magnitude −0.81 for observers on Earth (whoever they may be). On the other hand, if we could time-travel to more than a few million years in the past or more than a few million years in the future, we would find Vega so distant as not to be visible to the naked eye.

A more famous fact about Vega is what happens to its position in our sky over the course of 25,800 years. This is the period of the vastly slow (to us) wobble of Earth's rotation axis called precession. During the course of this period, the north end of Earth's imaginary rotation axis describes a huge circle in the heavens and brings us a long succession of North Stars. We are currently only about a century away from the closest approach of the "north celestial pole" to the 2nd-magnitude star in Ursa Minor we call Polaris. But on the other side of the processional circle—about 13,000 years in the past and about 13,000 years in the future—the bright star closest to the north celestial pole is Vega. It is not very close to the pole at those times, but Vega is much brighter than any other star that earns the title of North Star in the 25,800-year cycle.

Vega in Legend and Lore

The ancient Greeks and Romans imagined that Lyra was a lyre—a harplike musical instrument. One nickname of Vega used by the ancient Romans has been translated "the Harp-Star." Lyra is most often supposed to represent the lyre of Orpheus, the greatest musician of classical mythology. This lyre was invented by the Greek god Hermes (whom the Romans called Mercury) from an old tortoise shell and seven cow-gut strings. Hermes gave it to Apollo, who in turn gave it to Orpheus. We haven't time to retell the adventures of Orpheus here (but we can note that he was a fellow Argonaut with Heracles, Kastor, and Polydeukes, who as Hercules and the Gemini twins have constel-

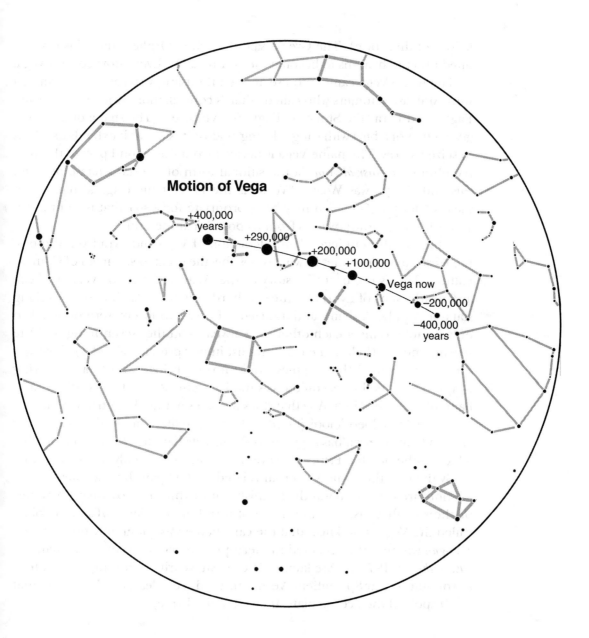

Motion of Vega

+400,000 years

+290,000

+200,000

+100,000

Vega now

−200,000

−400,000 years

The path of Vega through the heavens over a long period of time (note that the background stars would move, too—though not as far through the sky as our near neighbor Vega).

lations of their own). The Greeks said that when Orpheus died, his lyre was lifted into the heavens to become the constellation Lyra, adorned with Vega.

The name Vega came into use about a thousand years ago. It originated with Arab astronomers who called Altair's constellation, Aquila, "the Flying Eagle" and Lyra "the Stooping Eagle (or Vulture)." The shape of Lyra suggested that of a bird with wings closing against its sides as it extends its talons to seize its prey. The name Vega is derived from the second part of the Arabic phrase *an-Nasr al-Waqi* (a transitional form of the star's name between this and "Vega" was Wega). "Vega" is the part of the original name that means "stooping"—which may be appropriate for a star that is so bright it seems to stoop down over us from its position at the zenith.

A totally different legend and depiction of Vega and Lyra from those of the Western world comes in what is possibly the greatest star myth of Chinese culture. This is the beautiful story of the Weaving Princess (Vega, or Vega and nearby stars of Lyra) and the Cowherd (Altair, or Altair and its flanking stars in Aquila, Alshain, and Tarazed). These two lovers were doomed to only be able to meet each other once a year—on the "seventh night of the seventh moon" (held to be early August in our present calendar). That was the night when all the magpies would form a bridge with their fluttering wings so that the lovers could cross the Heavenly River of Stars—the bright band of Summer Milky Way that does indeed separate Vega from Altair.

To me, Lyra doesn't look just like a lyre or just like a rattle (the latter being an idea Guy Ottewell has suggested—Vega is its handle). Lyra also looks like a loom—the one the Princess weaves her glorious heavenly tapestries upon.

By the way, the name Vega is unrelated to the Spanish vega (meadow). I am not sure whether it was the Spanish word (which is also a Spanish name) or the star that inspired the name of a rather popular 1970s automobile called the Vega. I do know that the car's name was commonly pronounced *VAY-guh* whereas the accepted modern pronunciation of the star's name is *VEE-guh*. In 1977 at Stellafane, the great yearly gathering of amateur astronomers near Springfield, Vermont, I did see at least one Vega (car, that is). It sported the license plate ALYRAE (Alpha Lyrae).

The Possible Planet of Vega

Astronomers have no evidence that a weaving maiden lives in the Vega system. But there is now evidence of a planet circling this star.

The fact that Vega is surrounded by a shell of dust at a great distance was discovered in the early 1980s by the satellite IRAS, which imaged the universe at infrared wavelengths. This was the first evidence that solid matter orbits stars other than our Sun and led to stars with far-infrared excesses being called "Vega-like" objects. Could the dust be the dust component of a belt of comets out beyond any planets Vega might possess?

In 2002, studies of Vega at millimeter wavelengths revealed knots of dust at 60 and 75 AU. (about 5.6 and 7 billion miles) from Vega. Scientists interpret this as being caused by a Jupiter-like planet (though perhaps much bigger than Jupiter) in an eccentric (elongated) orbit that averages about 30 AU from Vega (similar to Neptune's distance from the Sun). Thus Vega became the first of the very bright stars in our sky for which there is evidence of an accompanying planet. Ironically, Vega was the source of the first communication from extraterrestrial intelligence in Carl Sagan's 1985 science-fiction novel *Contact*, later made into a movie.

What is the distance from Earth to Vega and its possible planet? A 1994 study found a distance of 25.1 plus or minus 0.15 light-years. A few years later, the Hipparcos survey judged the distance to Vega at 25.3 plus or minus 0.1 light-year. By a remarkable coincidence, these measurements place Vega at essentially the same distance from Earth as Fomalhaut, which IRAS also found to have a surrounding dust disk. Studies of Fomalhaut's circumstellar dust disk in 2003 suggested that it, too, is accompanied by a planet.

The Flattened, Hot-Poled Vega

On the night of July 16–17, 1857, Vega became the first star to have its photograph taken. That hundred-second exposure by the daguerreotype process was made at Harvard. In the decades that followed, studies established Vega as the "standard star" against which to measure others for magnitude, luminosity, temperature, color, and spectrum. Vega shines at an apparent magnitude of +0.03 and on the B-V index of colors is 0.0. It is an A0 main sequence star of Sirian type (another link between Vega and Sirius).

Astronomers eventually found, however, that Vega is actually a bit hotter and brighter than a star of its spectral class should be. What could explain this? In 1994, a group of researchers concluded that the model that would best fit the data would be one in which the rotational axis of Vega was pointing to within 5° of Earth and the star was rotating not in the five days previously estimated but a very rapid eleven hours. This rapid rotation would make Vega bulge at its equator and thus appear larger to us, if the star's pole was pointed toward us. The polar region facing us would be hotter and brighter than the equatorial region (because it is close to Vega's core). Matter at Vega's equator would be moving at more than 200 km/sec, with some of it possibly escaping from the star.

In 2005, these basic ideas about Vega's rotation rate, shape, polar orientation, and temperature were confirmed and greatly refined by a new study. The instrument that made this possible was Georgia State University's Center for High Angular Resolution Astronomy (CHARA) Array. The CHARA Array is a Y-shaped arrangement of 1-meter telescopes on Mount Wilson in California, which are linked together to achieve resolution as fine as 200 micro-arc seconds. A group headed by Jason Aufdenberg used the CHARA Array to determine that Vega rotates every 12.4 hours, which is 92 percent of its calculated breakup speed. The equatorial diameter of Vega is 2¾ solar diameters, but the polar diameter is only 2¼. As a result, Vega's poles must have a surface temperature of about 9,900° C (18,000° F), a huge 2,200° C (4,000° F) hotter than the equator. Since Vega points its pole toward us, we see it appear brighter than a standard A0 star. But if we factor in its shape and cooler equator, Vega should be emitting about thirty-seven times more light than the Sun—similar to other A0 stars.

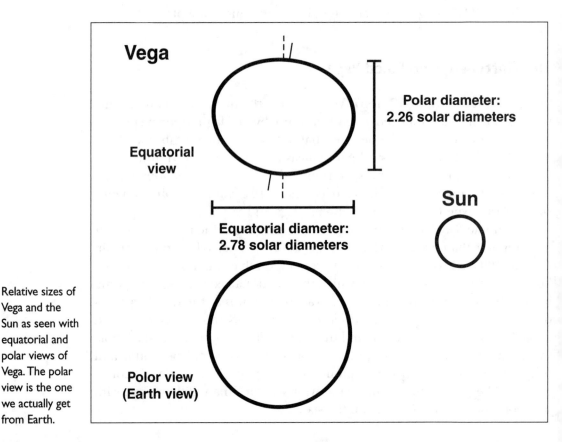

Relative sizes of Vega and the Sun as seen with equatorial and polar views of Vega. The polar view is the one we actually get from Earth.

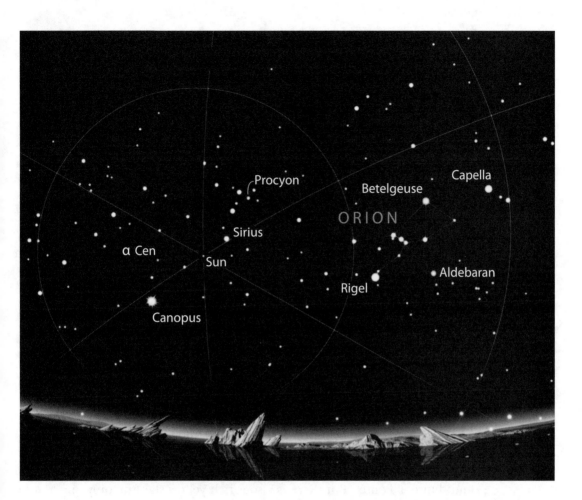

Vega is our most luminous neighbor until we get as far out as Arcturus at 36.7 and Capella at 42 light-years.

A view of the polar sky from Vega (or a planet orbiting upright in Vega's equatorial plane).

The Pole Star of Vega

What is the view like from Vega? The closest known stellar neighbor to Vega—a dim one—is 4.2 light-years from it. But most interesting are the distances of Earth's 1st-magnitude stars from Vega. Altair is 14.8 light-years from Vega—only a little closer to Vega than it is to us. That's also true of Arcturus, which is 32 light-years from Vega and a negative-magnitude object in Vega's sky. Sirius and Procyon, respectively, are 33 and 34 light-years from Vega, so they appear as modest 2nd- and 3rd-magnitude stars.

But there is an aspect of Vega's sky that we would find most interesting of all. For there is a corollary to the aforementioned fact that Vega points a rotation axis at us: our Sun is a pole star for Vega.

If Vega has an inhabitable planet, an observer on it would only see our Sun near the celestial pole if the planet were orbiting close to the plane of Vega's equator and had a low inclination of its axis to its orbital plane. But both of these assumptions are reasonable ones. We expect planetary systems to form in the equatorial planes of their parent stars. And in our own solar system, Mercury, Venus, and Jupiter have their rotation axes nearly perpendicular to their orbital plane.

Our Sun would appear as a modest magnitude 4.2 star in the sky of Vega. Not far to one side of it would be Sirius. And somewhat farther to the other side of it would be Canopus, shining only very slightly fainter than it does in Earth's sky.

13 CAPELLA

The palette of the bright stars of winter includes the many-coloredness of Sirius when twinkling down low, the blue-white of Rigel, the deep orange of Betelgeuse, the light orange of Aldebaran, and the subtle hues of Procyon and dimmer Pollux. But there is also a rich yellow star, the only one among the one hundred brightest nighttime stars that is overwhelmingly of G spectral class—the class of the Sun. This star is Capella—the sixth brightest in all the heavens. As we'll see, it is also the northernmost of all 1st-magnitude and brighter stars and consequently a star of three full seasons for most of the world's population.

Capella's First-in-Fall Beauty

Capella is the fourth brightest star visible from 40° N and is only marginally less bright than Arcturus and Vega. But those two stars are by far the brightest of the spring and summer constellations, respectively. In winter, Capella has the misfortune of having to share the sky with peerlessly brilliant Sirius. It is the second-brightest star in the bright winter constellations. But the slightly less bright Rigel and Betelgeuse shine in stunning Orion, Aldebaran

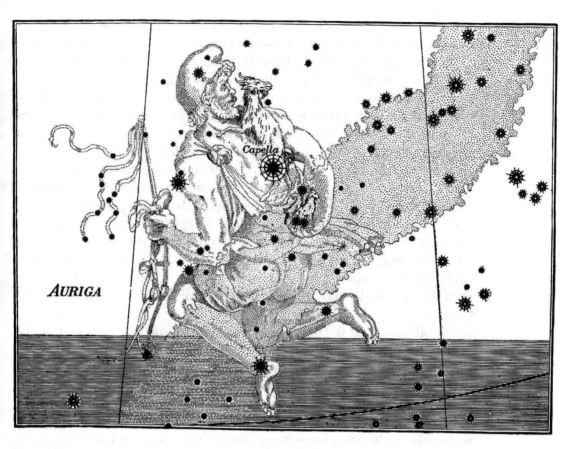

AURIGA

Capella

Auriga and
Capella.

is strikingly placed as the eye of Taurus with the Hyades cluster, and even
much dimmer Pollux may draw more attention than Capella because of its
close pairing with Castor as the "twin" stars of Gemini, the Twins. Capella
beams from out of a less well-known, less mythologically rich constellation,
Auriga, the Charioteer. The fact that Capella passes directly overhead for
observers near 46° N latitude—Portland (Oregon), Minneapolis-Saint Paul,
Montreal, much of Maine, much of Europe—is a notable fact. But even that
may work against Capella in the midst of winter evenings because observers
at most midnorthern latitudes have to crane their neck to see it then, need
to make a special effort to leave the more famous bright stars and constellations that are less far above typical eye level.

The saving grace for Capella is the way its northerliness brings it into view
long before the other bright winter stars and constellations appear. This
allows it to dominate the northeast sky for much of the evening in much of
autumn. It also permits Capella to linger late in the spring at nightfall, as

part of an arch composed of Pollux and Castor and Procyon—all stars a lot less northerly than Capella but also a lot farther east on the celestial sphere.

That first-in-fall visibility of Capella is what is really special. As a matter of fact, the initial sight many amateur astronomers get of Capella after its late spring disappearance into the Sun's afterglow is not in early autumn but in August. The sight happens as part of skygazers looking for Perseid meteors. Many Perseid observers begin their serious meteor watch around the midnight hour—and that is just when Capella's climb up the northeast is occurring, accompanying the exciting increase in numbers of meteors. As autumn and the first really chilly nights come to midnorthern latitudes, the rising of Capella occurs earlier and earlier—but nightfalls are also getting earlier and earlier at that time of year, so the process of Capella rising earlier relative to the night is slow. One thing that remains the same is the order of bright sights climbing the northeast: first the zigzag pattern of Cassiopeia, then straight below it noble Perseus, and then below Perseus—and far brighter than the 2nd- and 3rd-magnitude stars of Cassiopeia and Perseus—mighty Capella.

We often say that the only 1st-magnitude (or brighter) star of autumn is comparatively dim Fomalhaut, which arcs rather low across the southern sky for viewers at midnorthern latitudes. But as the great twentieth-century observer Leslie Peltier noted on his farm in Ohio, Capella is so much farther north than Fomalhaut that the two stars are "co-risers" (they rise at about the same time) at around 40° N latitude. This is true even though Capella is over six hours of right ascension—one-quarter of the way around the circle of the whole celestial sphere—farther east than Fomalhaut. For observers around 40° N latitude, the northernmost of the brightest stars (Capella) rises at the same time as the southernmost visible from that latitude (Fomalhaut), even though the former is at least one full season in RA behind the latter. Fomalhaut is sometimes called the Autumn Star, but if you look northeast, the title could also be accorded to the much brighter Capella.

Capella for Christmas

Autumn passes. Where is Capella by Christmastime? As darkness begins to fall around 40° N, Capella is still only about 30° up the northeast sky. That's far enough up for a good, practically undimmed view. But, more notably, it is a lot higher than fainter Aldebaran, and only one other of the remaining six 1st-magnitude or brighter stars of winter is yet above the horizon—Betelgeuse, which is literally on the horizon (so really too dimmed by atmospheric extinction and any haze to be seen yet). Interestingly, this occurs just before

0 hours sidereal time, and that means that Fomalhaut has recently passed the meridian in the due south and is 30° high just like Capella. But Fomalhaut will never get any higher for observers at 40° N, whereas Capella will climb until it passes near (just north of) the sky's summit—about 11:00 P.M.

Another interesting point is that despite the fact Capella is one-third the way up the sky at nightfall at Christmas, Rigel has not quite yet risen. This is interesting because Rigel—ever so slightly fainter than Capella—happens to lie at almost exactly the same right ascension as Capella (5h 15m compared to Capella's 5h 17m). Capella is almost precisely due north of Rigel. The reason it is already well up when Rigel has yet to rise for people in midnorthern lands is that Capella lies a huge 54° north of Rigel. By 11:00 P.M. Rigel will have already traveled its much shorter arc across the sky to the meridian and will be crossing the meridian when Capella is.

I have a very special attraction to where Capella is at Christmas because I have an important and very fond memory of what I saw at nightfall on December 25 back in 1967. That day I had received my first really useful astronomical telescope as a Christmas present, just a few weeks after my thirteenth birthday. The telescope was a 4 ¼-inch Edmund reflector, a popular telescope for many serious young stargazers in those times. The day was cold but crystal clear, and I knew the forecast was such that I could expect to be able to use the telescope that evening. I could scarcely wait for nightfall. When it came, there was no Moon, and initially the few planets and all the really bright stars were hidden behind surrounding trees—except for one ascending the northeast sky: Capella. Capella was the first star—indeed the first celestial object—I ever looked at through my first legitimate astronomical telescope.

Astronomers call the experience of the initial observation with a new telescope by a special name: "first light." It is one of the most uplifting of all events, whether the instrument is a world-class observatory telescope or a tiny backyard scope. The light of my life's most important "first light" moment was that of Capella.

A Star for Three Seasons

Another memorable observation I made with Capella in my view came on a winter night when the star and its constellation Auriga were near the zenith. I was directly shaded from a modest nearby streetlight and looking straight up at a clear, dark view of Capella and Auriga, when suddenly I started seeing shining specks of glitter slowly drifting down to my upturned face. Despite my ability to perceive stars of 5th magnitude or dimmer, there were snowflakes falling. They seemed like a shower of sparkling stardust or actual stars gently floating down upon me.

As noted earlier in this chapter, however, Capella is a star of spring, too. No one has captured better its role as a natural presence of three entire seasons—autumn, winter, and spring—than Martha Evans Martin in her book *The Friendly Stars*. The first chapter devoted to an individual star in Martin's classic is about Capella. One of its paragraphs states:

> When you watch the birds congregating in noisy flocks in the morning for the fall migration, and in the afternoon look for the first fringed gentians, look for Capella in the northeastern sky in the evening. When the trees are bare and the berries are wrapped in ice and snow, so that the winter birds greedily gather in what bounty you throw to them, you will find Capella shining almost directly overhead early in the evening. During the spring months, when the air is full of the stir of the awakening earth, and other stars are demanding our attention to their return in the east after long absences, Capella is hurrying on towards the northwest, no longer charming us with its novelty, but still as bright and fair as ever and ready to fill its place in the brilliant gathering of the stars of spring.

Summer Capella and "the Clouds of Capella"

Actually, Capella is a star for *all* seasons—if you live or visit somewhere north of about 45° N latitude. On the night of July 5–6, 1982, I was in Parshall, North Dakota (latitude about 48° N). That night, I was observing the longest total eclipse of the Moon in U.S. history (the total eclipse lasted about 106 minutes)—on the shortest night I'd ever experienced. One of the highlights happened at the end of the total eclipse. The bright sliver of Moon started to come back out in the south sky and was immediately greeted in the facing north sky by a sudden flaring cheerful fanfare of celebratory Northern Lights. But it was in the early stages of the eclipse that I and my fellow observers enjoyed the sight of a strange bright star hovering low above the due north horizon—Capella!

I wrote about this Capella experience in my "North Hemisphere's Sky" column in the January 2007 issue of *Sky & Telescope*. I got a response about the column from a well-known and veteran Canadian observer, Alister Ling. Ling noted an interesting connection between Capella and one of our atmosphere's most mysterious phenomena—*noctilucent clouds* (NLCs).

The word *nocitlucent* is Latin for "night-shining," for these are the highest of all clouds in Earth's atmosphere and are therefore still lit by the Sun when it is as much as 16 degrees below the horizon as seen from Earth's surface—almost full night. They appear as delicate silver-blue formations, sometimes with a golden lower edge. By far the best chance to view them comes in summer. That's when the mesopause in our atmosphere is actually at its coldest and when long twilights provide hours in which NLCs may be spotted—in the vicinity of summer's low Capella. Ling wrote: "Just thought you should know that a bunch of us in Edmonton, the prime NLC latitude in North America (can't speak for the Europeans), have a fondness for Capella. In trying to encourage more observers of NLC, Mark Zalcik lyrically refers to NLC as 'the clouds of Capella,' since 90 percent of NLC images have Capella in them."

But Ling also mentioned another distinction of Capella for observers at his latitude: "Capella is also our prime false UFO candidate, because in summer it sits above the northern horizon in twilight when folk are out on late evening walks, [and see it] twinkling madly away."

Capella's Name and Lore

The name Capella is sometimes translated as Latin for "she-goat," but that would be "Capra." Capella is the diminutive form of Capra—thus "little she-goat." But a final twist on this comes from the great star-name authority Paul Kunitzsch. He suggests that cases like this "are likely meant to indicate an atypical use of words, as an animal's name given to a star instead of to the animal (rather than meaning 'small' animals or personages)."

Whatever the exact form of its name, Capella certainly has been imagined to be a goat from at least the times of ancient Greece (who called Capella "Aix"). In connection with this, a small triangle of 3rd- and 4th-magnitude stars 3° to 5° southwest of Capella have been known as "the Kids" (baby goats). The most interesting of the Kids is Epsilon Aurigae (the Kid closest to Capella in the sky). Epsilon Aurigae lies about 3,000 light-years from Earth (compared to 42 light-years for Capella) and therefore has an absolute magnitude of about −7. What's most unusual of all about Epsilon Aurigae is its eclipses. These eclipses occur twenty-seven years apart! An eclipse begins with the star's dimming slowly (over the course of months) from about magnitude 3.0 to 3.8. Epsilon is then at the minimum brightness for about a year—before beginning a similarly slow recovery to its original brightness. These strange eclipses were once theorized to be caused by a star so huge it would fill our solar system out to the orbit of Uranus. Now we believe that the eclipsing companion is surrounded by a vast disk of interstellar dust. Whatever the cause of the eclipses, you're in luck if you are reading this book before 2009 to 2011—because that is the time of the next eclipse.

It is not known why Capella and the Kids are being held by the Charioteer that is formed by the overall constellation Auriga. Drawings of the Charioteer over the past two thousand years have shown him with the mother goat Capella and her kids on his left shoulder. The goat and its children seemed to have been regarded as a separate constellation from the Charioteer in Greece before the classical period. The goat in Greek mythology most often linked with Capella is one called Amalthea, or one owned by the nymph Amalthea. This was the goat whose milk nourished baby Zeus.

Some ancient Greek writers had a related but alternative identification of Capella. They claimed the star represented a horn of the goat that baby Zeus had broken off in play and which was placed in the heavens as the Horn of Plenty, the Cornucopia. According to one version of the legend, the Cornucopia would fill up with an endless supply not just of food but of whatever one wished for. (By the way, note that the name of the constellation Capricornus means "horned goat," the prancing of little goats is "capering" and "capricious," and our modern English word *copious* means "plentiful.")

It has been argued that Capella is mentioned in an Akkadian inscription that dates back to about 2000 BC. One thing for sure is that Capella's place on the famous Denderah zodiac is marked, as R. H. Allen notes, by "a mummified cat in the outstretched hand of a male figure crowned with feathers." Allen also reports the claim that Capella was an important star in the temples of the ancient Egyptian god Ptah, the Opener, and may even have had the name Ptah applied to it.

The ancient Greeks and other cultures associated Capella with the rainy season. It was called "the rainy Goat-starre." And Aratos wrote that "Capella's course admiring landsmen trace,/But sailors hate her inauspicious face." Capella has also been called "the Shepherds' Star" by English poets and was connected with shepherds by the ancient Peruvians, the Quichuas, who knew Capella as Colca.

Similar ideas about Capella as a herder or watcher of a flock occurred among the early Arabs. One name they gave to Capella was Al Rakib. That meant "the Driver" because, says R. H. Allen, "lying far to the north, it was prominent in the evening sky before other stars became visible, and so apparently watching over them." Allen may simply mean visible first as twilight faded, but I suspect the idea may have been what I and many other modern observers have noted—that Capella appears first, impressively in advance of the great winter flock of constellations. But Allen notes another of the early Arab names for Capella was Al Hadi, "the Singer" who cheers on a troop of camels, which in this case would be the Pleiades. The Pleiades rose at about the same time as Capella for Arabia then, just a little while after Capella for observers around 40° N now.

In ancient India, Capella was considered a sacred star, for it was there called Brahma Ridaya—"the Heart of Brahma." In the nineteenth century, Alfred, Lord Tennyson referred to Capella as "a glorious crown" in his poem "Maud." In modern times, the *Star Trek* TV episode "Friday's Child" was set on the planet Capella IV.

Here's a final note on the name of Capella. It may not take long after you learn of Capella before you wonder if there is any relation between the name of the star and the term for the style of singing without instrumental accompaniment. But the latter, though often misspelled "a capella," is actually "a cappella." And it is neither from Latin nor is related to any word for goat. Actually, *a cappella* is Italian for "like in the chapel"—no doubt a reference to the long tradition of church choirs singing with no music.

Double Capella

We've discussed observing Capella in the context of the sky and the seasons, and its name and lore in the context of history and story. Now let's turn to the topic of Capella's nature in space. In this respect, it turns out to be one of the most distinctive and interesting of all the brightest stars.

First, the vast majority of the point of light we call Capella is coming from not one but two stars. That's not unusual among the brightest stars. More distinctive (but still not unique among the brilliant stars) is how similar Capella A and B are to each other. They are truly twins in size—each about ten times the width of the Sun. They orbit each other at a distance of about 60 million miles apart (closer together than the Sun and Venus) in a period of only 104 days. This helps us calculate that the mass of each of the Capella pair must be approximately 2.5 times that of the Sun—but the brighter (Capella A) must be a little more massive. Capella A shines with eighty times the luminosity of the Sun, B with fifty. At Capella's distance of 42 light-years, that works out to magnitude 0.6 and 1.1 for A and B, respectively. The composite magnitude of the two is +0.08 (only 0.05 dimmer than Vega), making it the sixth brightest of all night's stars. But if you already know a little bit about astronomy, you may know that Capella A and B do not make a visual

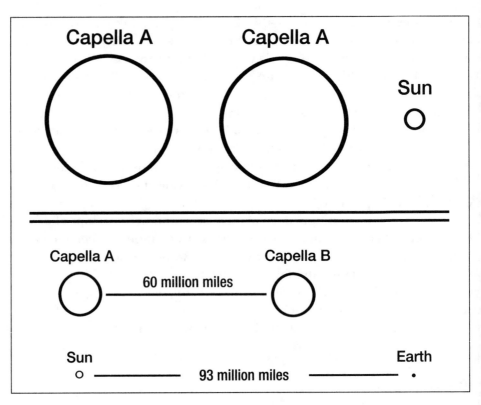

Relative sizes of Capella A, Capella B, and the Sun (top diagram); relative sizes and separations of Capella A and B versus the Sun and Earth (Earth not drawn to scale—it would be much smaller).

pairing—the duplicity of Capella was discovered spectroscopically. That is common among the brightest stars. But what is unique about Capella in this respect is how close it is to being visually splittable. The separation of Capella A and B is 0.04″, which theoretically could be split with about a 12-inch telescope. The problem is that the seeing in Earth's unsteady atmosphere is almost never good enough to allow such a split anywhere in the world. Still, I have to wonder whether Capella could just be split by a large telescope at the best locations on Earth on the best nights. Astronomers have been able to obtain photographs of the pair separated from each other by using interferometry.

Like so many of our bright stars, Capella also includes a distant red-dwarf member in its system—in this case, a 10th-magnitude object located about 12′ to the southeast of Capella AB. It is about 11,000 AU away but shares a common proper motion through space. Since the letters C through G were assigned to faint unrelated field stars, this red dwarf is commonly known as Capella H. Capella H, however, turns out to be really a rather close pair of red dwarfs, Ha and Hb. These two are, respectively, about one-half and one-quarter the diameter of the Sun, but even their combined light is only about

1 percent of the luminosity of the Sun. They orbit each other in a period of 388 years.

Yellow Capella

Now we come to what is most interesting about Capella A and B as suns: they are both of G spectral type, like our own Sun. There is no other pair of stars of which this is true among the hundred brightest as seen from Earth. In the case of Capella, it is difficult to establish the subtypes of the component suns from the separate spectrum of each of the couple. But one expert has A and B as a G8 and a G0 star, another as a G6 and a G2. G2 is the exact spectral subtype of the Sun. But the Sun is a luminosity class V star, a main sequence yellow dwarf. Both Capella A and B are in luminosity class III, which makes them yellow giants. Both have ceased burning hydrogen in their cores and are expanding to become red giants within the next few million years. According to star expert James Kaler, the slightly more massive (and therefore more evolved) star, Capella A, has very likely begun converting helium to carbon within itself. Capella B seems to have a contracting helium core that has not yet fired up.

Past Capella

What about Capella in the past? Both Capella A and B were probably spectral type A when they were still on the main sequence. Both are similar in mass to Vega, an A-class star, so perhaps these two were once somewhat like twin Vegas.

If we look at approximately the past one million years and the next million, we find that five well-known bright stars of today are at different times the brightest star in Earth's sky. The stars are Sirius, Canopus, Vega, Aldebaran, and Capella. Capella ruled for the shortest period of any of these five: the 50,000 years between −210,000 and −160,000 years. It was Earth's brightest star after Aldebaran and before Canopus (which was then followed by the current reign of Sirius that we live in). Interestingly, Capella had its greatest apparent brightness before its reign as the brightest star. It reached its peak magnitude from Earth 240,000 years ago when it came to within 28 light-years of Earth (compared to 42 light-years today).

Capella has a proper motion very similar to stars of a Taurus Moving Group associated with the Hyades, and there has been speculation that Capella could be an outlying member of the moving group. But its encounter with Aldebaran (see chapter 21) a few hundred thousand years ago is far more interesting.

The View from Capella

From near Capella A and B, an observer would see Ha and Hb as a single orange point of light shining somewhat brighter than Sirius does as seen from Earth. But an observer in the Ha/Hb vicinity would get a truly majestic view of A and B. The bright pair would appear to the unaided (human) eye as a golden point (or mass) of radiance almost as bright as a Full Moon here on Earth.

What about the view of other star systems from Capella? At least five of them lie within 10 light-years of Capella, but the only relatively bright one of these is Lambda Aurigae—located about 4.4 light-years away. Pollux and Castor are brighter in Capella's sky than they are in our own. When we look up at Gemini and see Pollux and Castor only 4 ½° apart in the sky, we aren't surprised to learn that Pollux is 23 light-years from Capella and Castor 26 light-years away. But at first it's a bit of a jolt to hear that Pollux and Castor are themselves 18 light-years apart. We tend to forget that Castor is a lot farther from Earth than Pollux (52 light-years compared to Pollux's 33.5). Capella is intermediate in distance from us compared to them.

An Exchange of Side-by-Side Champions

One final fascinating twist involves Capella and the 2nd-magnitude star with which it forms the north end of Auriga. Only about 7½° from Capella, 1.9-magnitude Beta Aurigae is better known as Menkalinan. Amazingly, whereas Capella is drifting away from us after having been 28 light-years distant and Earth's brightest star those few hundred thousand years ago, Menkalinan is going to become our brightest star and also come to a minimum distance of about 28 light-years from Earth—more than a million years from now.

The current distance of Menkalinan is 82 light-years. After almost 2 million years of our current 1st-magnitude stars Canopus, Aldebaran, Capella, Sirius, and Vega having exchanged the title of brightest star, the far less famous Menkalinan will take over the title from about 990,000 to 1,150,000 years in the future. At the end of that period, Menkalinan will be outshined by what is now a mere magnitude 4.7 star, Delta Scuti (Delta Scuti is currently 187 light-years away but will pass just 9 light-years from Earth 1,250,000 years in the future). But Menkalinan will continue nearing us after that, reaching its 28 light-year closest approach around 1,190,000 years hence.

How remarkable it is that two stars next to each other in a constellation are a receding past champion in brightness and an approaching future champion. Within the span of 10 million years that has been studied, there is no other case anything like this in all the heavens.

RIGEL

E ven though it is designated Beta Orionis (Betelgeuse is Alpha), Rigel is almost always the brightest star of the brightest constellation. It is the sapphire to contrast with Betelgeuse's topaz. Rigel is the absolutely classic "blue giant" star. It is one of the few most radiant flames on our section of our galaxy's spiral arm. It is unquestionably the most luminous star within a thousand light-years of Earth. Rigel is so important it has another star for a "footstool." It may be the source of illumination for a "witch's head." And it definitely is the bright fire at the start of the amazingly long river constellation of the sky.

In the noblest of all constellations, Rigel must be the noblest of all individual objects to the naked eye.

Rating Rigel

With an apparent magnitude of +0.12, Rigel shines ever so marginally dimmer than Capella and only a tiny bit dimmer than Arcturus and Vega. When careful measurements are made, Rigel turns out to be the fifth brightest star visible from midnorthern latitudes and the seventh brightest of all stars in the heavens.

As described in my book *The 50 Best Sights in Astronomy*, Rigel was the first star that I ever tried to identify. It was initially quite believable to me that this was blue-white Sirius, brightest of stars—until I saw the even more splendid Sirius. But although Sirius appears much brighter in our sky, Rigel is very

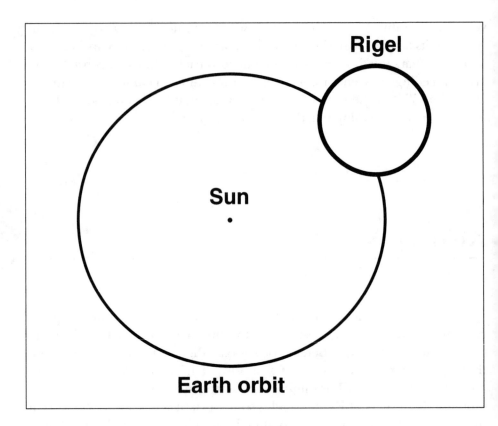

Rigel

Sun
•

Earth orbit

Relative sizes
of Rigel and
Earth's orbit.

roughly a hundred times farther away than Sirius. If it were as close to Earth as Sirius is, Rigel would rival a half Moon in brightness in our sky.

Rigel is classified as a star of luminosity class I, which means it is a "supergiant." It is the most luminous of all the very bright stars in our sky save for Deneb (both are usually considered to be in luminosity class Ia). But Rigel is roughly two to three times closer to us than Deneb is, so it easily outshines that star in apparent magnitude.

Rigel has long been considered the most luminous sun in the spectacular Orion Association, the immense flock of hot, brilliant stars of O and B spectral class that are mostly responsible for the brilliance of Orion and some of its neighboring sky. But a few decades ago the distance of the star was estimated to be about 900 light-years. In recent years, we normally see the distance Hipparcos determined for Rigel—about 775 light-years. It's important to remember that, this far out, the Hipparcos data would have an average margin of error of about 20 percent.

If Rigel's distance happened to be exactly 775 light-years, its absolute magnitude would be −6.8 and its luminosity equal to forty thousand suns.

But Rigel is a B8 star, extremely hot (surface temperature about 11,000 K), and therefore a mighty producer of ultraviolet radiation. If we add this UV into the calculation, Rigel would be releasing about sixty-six thousand times as much energy as our Sun.

Rigel as a Sun

What are the size and mass of Rigel? Although blue supergiants like Rigel are not as large as red supergiants, they are still enormous. Rigel is estimated to be about seventy times wider than our Sun. That's a little larger than Canopus and surpassed in size only by Antares and Betelgeuse among the 1st-magnitude stars.

Less than three decades ago, Robert Burnham Jr. quoted a mass of possibly fifty times that of the Sun for Rigel. We now think the real value is much smaller—Rigel has (or at least started out with) about seventeen times the mass of the Sun. Only Deneb may be more massive among the 1st-magnitude stars. But the prodigal young star Rigel is blowing off its matter in a big way. According to James Kaler, Rigel may lose enough mass in its mere few million years of life to evade the fate of going supernova. Just possibly, it will lose just enough mass to collapse into a rare heavy oxygen-neon white dwarf. Rigel has probably already reached the stage of converting its helium into carbon and oxygen.

Rigel varies slightly in brightness. Most sources cite variations of only about 0.03 magnitude, though a few say that fluctuations of as much as 0.3 magnitude are possible (they must be very rare). The average period of variations is supposedly about twenty-five days.

Rigel does not travel through space alone. But we'll get to that matter in a moment. First, let's discuss the name and the lore of Rigel.

The Foot of Al Jauzah

Rigel is sometimes said to mark the western knee of Orion, the Hunter, but its name suggests that it marks the foot (we can picture Orion with this forward foot upraised as he prepares to strike with his club at the onrushing Taurus, the Bull). Paul Kunitzsch tells us that the name Rigel was first used in the West more than a thousand years ago, when it was adopted from the Arabic title *rijl al jauza*—"the foot of al Jauzah." We do not know who al Jauzah was but the word seems to refer to Orion as a feminine figure who is perhaps "the one in the middle." I've discussed "the centralities" of Orion in my book *The 50 Best Sights in Astronomy*. It's possible the middle alluded to with regard to Orion is the constellation's position on the celestial equator or in the center of the crowd of the bright winter constellations.

To the medieval Arabs then, Rigel was the foot of al Jauzah—and that foot had a footstool composed of some far dimmer stars near Rigel in Orion and Eridanus. The title for "Forward Footstool" gave rise to the name Cursa for the brightest of these stars, Beta Eridani. Cursa is the magnitude 2.8 star that is first at the northwestern end in the long, mostly north–south pattern of Eridanus, the River.

Forming almost a right triangle with Rigel and Cursa is the photographically fascinating but visually very dim and difficult IC 2118—the Witch Head Nebula (named, as are most deep-sky objects, for its shape). The Witch Head seems to be an SNR—supernova remnant—which is currently shining by reflected light from the mighty radiance of Rigel.

Because Rigel is part of such a striking pattern of stars, it is the pattern—Orion—which has gotten most of the lore. Medieval Arab astronomer Al Sufi reported, however, that Rigel was also called Ra'i al Jauza, "the Herdsman of the Jauza," the herdsman's camels supposedly being Alpha Orionis (Betelgeuse), Gamma Orionis (Bellatrix), Kappa Orionis (Saiph), and Delta Orionis (Mintaka). But why would only one of the stars of Orion's Belt—Mintaka—be considered a camel in this picturing?

Several hundred years ago, some authorities gave Rigel such spellings as Regel and Riglon. Poets since then have called the star Algebar and Elgebar.

R. H. Allen claims that in "the Norsemen's astronomy" Rigel marked one of the big toes of Orwandil (Orion). The other big toe supposedly got frostbitten on a journey of the giant Orwandil, so Thor broke it off and flung it into the northern heavens. There it can still be seen as the 4th-magnitude Alcor, the little star that most naked eyes can glimpse right next to Mizar at the bend in the Big Dipper's handle.

The Companion of Rigel

Interestingly, whereas detecting Alcor is a good test of having satisfactory naked-eye vision, seeing the companion of Rigel is a good test for a 6-inch telescope. (James Mullaney reminds us, however, that under excellent conditions Rigel B may be glimpsed with a telescope with as small as 3 inches of aperture at 90×.) Rigel B is a lovely sight beside its brilliant big brother. A 10-inch or larger telescope shows it very prominently. Observing it is the first, easiest step in a progression that more experienced observers can

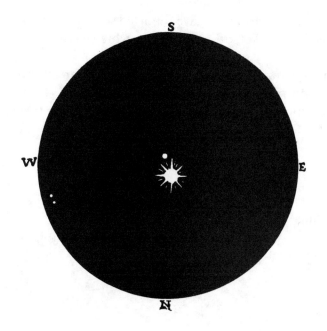

use on the way to trying see the companion of Sirius. First, see Rigel's companion. Then, if your instrument and the seeing conditions allow, move on to the harder task of detecting the companion of magnitude 1.5 Adhara (Epsilon Canis Majoris). Only if your atmosphere is steady enough for Adhara's companion to be very easily split should you try looking for Sirius B.

The companion of Rigel appears as a magnitude 6.8 object that is 9.5" from Rigel. Robert Burnham Jr. made the point that at Rigel's distance this very hot—spectral type B5—star is of similar temperature and luminosity to the powerful Regulus. Regulus is about 140 times more luminous than our Sun in visible light but probably lies about 10 times closer to us than Rigel and Rigel's companion. How mighty a star is Rigel that even its tremendously dimmer-looking companion is as radiant a star as Regulus!

Actually, however, things are a little more complex. As Burnham knew, the companion of Rigel turns out to be itself a close double star. The pair are in reality two magnitude 7.6 hot B-type stars that are only 0.1" apart. The duplicity was discovered spectroscopically, but there have been remarkable claims of the pair being split visually. Back in 1871, S. W. Burnham noted a distinct elongation of the companion's image in a 6-inch refractor and then apparently verified the observation with the 18½-inch telescope at Dearborn Observatory. Some famous and skilled observers—including the eagle-eyed E. E. Barnard—have reported seeing duplicity of elongation of the star's image. Yet other attempts which should have succeeded have not.

The pair of companions—Rigel B and C—are far removed from Rigel itself, at least fifty times the Sun–Pluto distance from it. There is little doubt, however, that these stars are physically bound to Rigel, traveling through space with it.

Rigel poised at the edge of Saturn, as viewed by the Cassini spacecraft orbiting Saturn.

PROCYON

15

Research of the past decade or so has revealed Procyon and its white dwarf companion to be distinctive suns of unusual interest. None of the other very bright stars in our sky are much at all like them. And learning more about them may greatly increase science's understanding of stellar evolution.

It's a good thing that Procyon and Procyon B are now found to be surprisingly special. Because otherwise, these stars have always seemed to fall short of any superlatives in relation to their neighbors in the sky and space. We should start our examination of Procyon with these many and famous ways in which it does not come in first place.

Suffering by Comparison

At an apparent magnitude of 0.34, Procyon is the sixth brightest star visible from midnorthern latitudes and the eighth brightest visible from anywhere on Earth. Yet its location in the heavens has doomed it to always get less attention than some of its neighbors. It always seems to be an also-ran in all the major categories of qualities for which we esteem stars. Procyon is outshined by three stars that are not all that far from it in the winter constellations. Neighbored by five bright and conspicuous constellations, Procyon's own constellation—Canis Minor, the Little Dog—is a faint patch of sky with only two other stars brighter than 5th magnitude.

But we always come back to the same point: all the other bright constellations or bright stars of winter are not what primarily overshadow Procyon—it's Sirius. The brightest of all night's stars shines not far from Procyon and is the great light of the magnificent, bright Canis Major, the Big Dog.

The Lesser Dog Star

For thousands of years, Sirius has been known as the glorious Dog Star, even credited (or rather blamed) with adding its heat to the Sun's to produce the "dog days of summer." Procyon has had to settle for, now and then, being

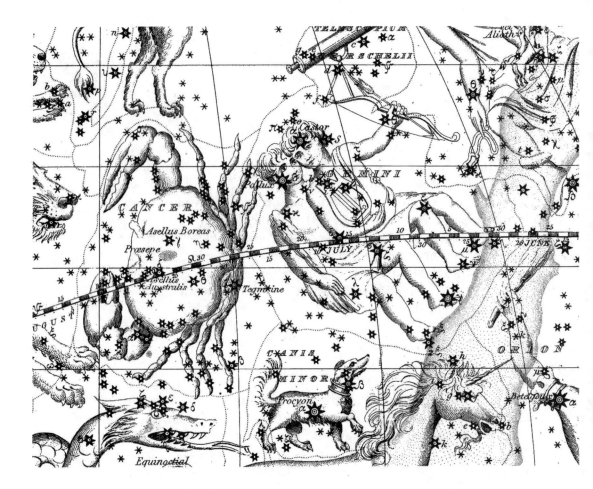

called the little or lesser Dog Star. A comparison between Sirius and Procyon finds the latter the lesser object in virtually every respect, not just in apparent magnitude.

Procyon shines almost 1¾ magnitudes dimmer in our sky than Sirius, almost five times fainter. Procyon is remarkable for being only 11.4 light-years from Earth. That makes it about the fourteenth-closest star system of any kind and the second-closest bright star visible to most of the world's population. But Sirius is the closest of all in that category. Since Sirius is closer to us (almost 3 light-years closer), we know that its true brightness must not be as much greater than Procyon's as it appears from Earth. But the absolute magnitude of Sirius is still 1.45 compared to 2.62 for Procyon. In other words, Sirius is about twenty-five times more luminous than our Sun, whereas Procyon is less than eight times brighter than our Sun.

Sirius is an A1 star with a surface temperature of about 9,880 K, while Procyon is an F5 star with a surface temperature of only 6,530 K (still considerably hotter than the Sun's). At least Procyon is larger than Sirius—about 2.1 solar diameters compared to about 1.75 for Sirius. But Sirius is heftier, possessing 2.1 times the mass of our Sun. Procyon is only about 1.4 or 1.5 times as massive as the Sun.

What about the comparison between the white-dwarf companions of Sirius and Procyon? Sirius B is a difficult sight for amateur astronomers with medium-size to large telescopes, even in the decades when it appears out around its maximum separation from its primary. But Procyon B is even fainter and much closer to its primary, and so has only been glimpsed with really huge telescopes, such as the world's largest refractors. Procyon B was once regarded as smaller and denser than Sirius B—which, for a white dwarf, would have been more impressive. We now believe, however, that Procyon B is larger, less dense, and cooler than Sirius B.

Fortunately, as we'll discuss in a moment, those properties of Procyon B are actually indicative of its being a fundamentally different type of white dwarf from Sirius B—a very interesting state of affairs. Before we examine closely the wonders of Procyon A and B as suns, however, let's consider more of the fascinating observational information and lore about Procyon.

Before the Dog

The name Procyon is pronounced *pro-SY-on*. And even the very name of this star is a bow to Sirius. It is ancient Greek—*pro-kuon*—for "before the dog," or, perhaps more fittingly, "the One Preceding the Dog" (Antecanis in Latin). The Dog is Sirius, and Procyon is its harbinger because, at some key latitudes and times in history, Procyon rises soon before Sirius, announcing it. One such time and place is our present time at midnorthern latitudes. Viewers near 40° N see Procyon come up about 10 to 15 minutes before Sirius (however, Procyon gains altitude more rapidly, so you may see it peek above your tree line more than 10 to 15 minutes before Sirius). This is true even though Procyon is almost an hour of right ascension farther east than Sirius, because it is also 22° of declination farther north. The more northerly location of Procyon makes its span of time above the horizon much greater than Sirius's for people at midnorthern latitudes. Although these observers see Procyon just before Sirius, they don't see Procyon set until about 2 hours after Sirius. That makes a major difference in how late in the spring the two stars remain visible. Sirius is glimpsed low in dusk until about early April; Procyon, until early June. At April and May nightfalls, Procyon lingers prominently as part of an arch of departing bright stars of winter: Procyon, Pollux and Castor, and Capella.

Procyon has some interesting positional connections of other sorts with other bright stars. Procyon is only about 6' of right ascension farther west than Pollux. Capella and Rigel are only 2' different in right ascension but are 54° apart in declination, with Capella very far above the eye level (or slightly elevated) stare that fixes on Rigel. By contrast, Procyon and Pollux are less than 23° away from each other. Procyon is about 27° from Castor, 26° from Betelgeuse, and—most important—26° from Sirius. Sirius itself is 27° from Betelgeuse. The pattern of Sirius, Betelgeuse, and Procyon is an almost perfect equilateral triangle. In the past few decades, this pattern has achieved considerable fame as "the Winter Triangle." As I said earlier in this book, I find the Winter Triangle interesting but no match for the Summer Triangle, despite its superior brightness to its warm-weather counterpart. The problem with the Winter Triangle is that Betelgeuse is too inextricably a part of the striking pattern of Orion and that other bright stars (such as Pollux and Castor) and constellations (for instance, Canis Major and Gemini) distract from this triangle.

Procyon's positional relation to Betelgeuse is interesting in its own right. Betelgeuse has a declination of +7.4°, Procyon of +5.2°. Procyon is somewhat less than 2 hours of right ascension almost due east of Betelgeuse. Procyon is the 1st-magnitude star that is closest to the celestial equator. The two other 1st-magnitude stars that are within 10° of the celestial equator are Rigel (declination −8.2°) and Altair (+8.9°). (Spica is located at −11.2° and Regulus at +12.0°.)

There are actually, at first glance, some interesting connections between Procyon and Altair. They are almost exactly halfway around the sky from each other (almost exactly 12 hours of right ascension). Since both are slightly north of the celestial equator, viewers around 40° N see Procyon about to set just after Altair rises. Procyon and Altair are the second and third closest bright stars visible from midnorthern latitudes (11.4 and 16.7 light-years from Earth), with absolute magnitudes of 2.6 and 2.2. Although they belong to different spectral types, and the surface of Altair is almost exactly 1,000 K hotter than Procyon's, their colors aren't too much different, defining the visual transition between yellow-white and white stars. Otherwise white Procyon's hint of yellow rates it a B-V index of +0.42. Altair's almost indetectable trace of yellow rates a B-V figure of +0.22. Procyon is slightly bigger than Altair but slightly less massive. Small differences in mass can make big differences in a star's life, though. Altair is a highly flattened star spinning at a rate of no more than once every 10 hours. Procyon is believed to have an equatorial rotation speed of perhaps as little as

3.2 km/sec and to take possibly as long as thirty-three days to complete one turn.

The Lore of Procyon

Almost every legend concerning Procyon involves Sirius. One major tale, according to Paul Kunitzsch, may have sprung from attempts by the medieval Arabs to explain names given to these two stars by earlier Arab sky-watchers. Kunitzsch notes that the name Gomeisa has been wrongly applied in modern times to Beta Canis Minoris but originally belonged to Procyon. According to Kunitzsch, this title—*al-ghumaisa*—means "the Little Bleary-eyed One." Kunitzsch says the reason for according this name to Procyon, and the name *al-abur*, "the One Having Crossed Over [a river, etc.]," for Sirius was forgotten and led to a new Arabic fable to explain the two stars. In the story, he says, Procyon and Sirius were sisters, and their brother was *suhail*—Canopus. Canopus was the suitor of *al-jauza* (the early Arabic feminine version of Orion). According to Kunitzsch (in Tim Smart's translation): "In coitus, *suhail* broke the spine of *al-jauza*, thus killing her, after which *suhail* fled south. He was followed by his sister *al-abur*, who 'crossed over' the Milky Way (where the two stars now lie in the southern sky). Meanwhile *suhail*'s second sister, *al-ghumaisa*, was left alone north of the Milky Way, weeping, until her 'eyes became bleary.'" According to R. H. Allen, other Arabic storytellers told that Suhail "only went a-wooing of al Jauzah, who not only refused him, but very unceremoniously kicked him to the southern heavens." Apparently in some versions, Sirius and Procyon were not sisters of Canopus and Sirius in any case was angrily chasing Canopus after the latter murdered Orion.

Allen says that an occasional English title of Procyon was "the Northern Sirius." He also states that "Euphratean scholars" identify Procyon as being, from ancient times, Kakkab Paldera, Pallika, or Palura, which supposedly means "the Star of the Crossing of the Water-dog." Presumably the water alluded to is again the winter Milky Way, which in really dark skies noticeably separates Procyon from Sirius.

Allen adds that Hervey Islanders regarded Procyon as their goddess Vena.

Star-lore author Julius D. W. Staal tells a delightful story relating the Dog Stars to Gemini. He notes that the overall shape of Gemini can be pictured as a rectangle. This can be imagined as a table at which Pollux and Castor are sitting, eating their meal. Procyon and Sirius (or their constellations?) are dogs waiting patiently to be tossed some of the scraps from the meal.

Staal even says that some of the crumbs from the table are stars of magnitude 5 or 6 that are scattered between Procyon and Gemini.

Procyon A and B as Suns

Now let us turn to the fascinating facts about what makes Procyon and its companion distinctive.

Procyon itself is a star in the process of leaving the main sequence. It is judged to be an F5IV-V star, meaning it seems to be in the act of becoming, or has just become, a subgiant star (stars in luminosity class V, remember, are main sequence and called "dwarfs"). Procyon is brighter than expected for its spectral class, which suggests that it may have just exhausted hydrogen fusion in its core and started on its way to becoming a cooler but huger and therefore brighter star. In the 1990s, studies of oscillations of Procyon led to better understanding of its structure and to estimation of a much younger age for it. Previous calculations had it that Procyon was probably about 3 billion years old (compared to about 4.8 billion for our Sun and perhaps only 250 million for Sirius). The new estimate is that Procyon is 1.7 billion plus or minus 0.3 billion years old. But Procyon is now on its way to becoming a giant. And it may take only 10 to 100 million years for it to grow to about 80 to 150 times its present diameter.

Procyon is a very slight variable of the BY Draconis type. It has a corona of the solar type, which apparently is as hot as about 1.6 million K.

What else is physically interesting about Procyon? The fact that it goes through space with a remarkable white dwarf.

The existence of Procyon B was suspected as far back as 1840, from variation in the proper motion of Procyon. But astronomer Arthur von Auwers did not publish a computed period of forty years for the suspected companion until 1861. And then for many years even such great observers as Otto Struve and S. W. Burnham searched for Procyon B in vain. Finally, in 1896, astronomer John M. Schaeberle, using the 36-inch refractor at Lick Observatory, became the first to sight the companion of Procyon. The orbit of Procyon B is tilted 31.9° to our line of sight. When first observed, the star was about 4" from Procyon. But it has a fairly elongated orbit and, at present—2008—is at periastron (closest to Procyon in space). Its separation is only 2.23" at the start of 2008. This compares with 8.04" for Sirius and Sirius B. Procyon B shines at magnitude 10.8, making it slightly more than 10 magnitudes fainter than Procyon (the difference in brightness of Sirius and Sirius B is almost exactly 10 magnitudes).

The average separation of Procyon and Procyon B in space is 15 AU, but the two get as close together as 9 AU and as far apart as 21 AU.

A 2002 study of Procyon B with the Hubble Space Telescope led to some remarkable new insights about not just this star but about white dwarfs in general. Procyon B turns out to be larger, cooler, and less massive than previously believed. It is about 10,900 miles in diameter, compared to about 7,300 miles for Sirius B. It is, nevertheless, less massive than Sirius B—a solar mass of about 0.60 compared to Sirius B's approximately 1 solar mass. Procyon B is therefore considerably less dense than Sirius. Procyon B's material probably weighs about one-third of a metric ton per cubic centimeter. But that's only about 20 percent the density of Sirius B. The surface temperature of Procyon B was once thought to be about 9,700 K, but the new data and model estimate 7,740 K. It turns out that there are two major types of white dwarfs and that Sirius B is one kind, Procyon B the other. Sirius B is an example of the kind that has a thin skin of pure hydrogen; Procyon B is the kind with a helium atmosphere, the hydrogen probably being lost to stellar winds.

Both Procyon and Procyon B have been interestingly affected by material from each other. Procyon B was, of course, the more massive of the two stars long ago. It probably weighed in at about 2.1 solar masses. But it therefore evolved faster. After about 1.3 billion years on the main sequence, it swelled into its giant phases. In those phases, it must have lost about three-quarters of its mass to stellar winds.

The View from Procyon

Few of our brightest stars have a sky as interesting as Procyon does. Our Sun would be seen as a magnitude 2½ star in southern Aquila. But there would be two great stellar lights in the night sky. One would be Procyon B, which would shine as a torchlike point of light an average of a few times brighter than Earth's Full Moon if you were very close to Procyon. The other light would, of course, be Sirius. Sirius lies 5.2 light-years from Procyon and therefore should shine at about magnitude –2½ as seen from the lesser Dog Star. There are at least twelve star systems within 10 light-years of Procyon and one of them is very close: "Luyten's Star" (L-789-6ABC) is just 1.2 light-years from Procyon. But this orange dwarf system is so dim—absolute magnitude just 14.63—that it would not quite be visible to the unaided human eye even from nearby Procyon. On the other hand, from Luyten's Star, Procyon would burn at about magnitude –7.

16 ACHERNAR

Of all the 1st-magnitude stars, Achernar is probably the least famous. It's one of the six of these stars that is not visible from midnorthern latitudes. But it doesn't belong to one of the pairs of far-south stars that are close to each other in a brilliant constellation (such as Alpha and Beta Centauri and Alpha and Beta Crucis). Nor is Achernar the second-brightest star in the heavens, like Canopus, which also belongs to a blazing constellation, Carina. Instead, Achernar is remotely located in a largely dim constellation and is one of the two loneliest of the 1st-magnitude stars (the other is Fomalhaut).

The underappreciated Achernar deserves a lot more attention. It is, first of all, the ninth-brightest star in the heavens, usually outshining Betelgeuse. It also marks the southern end of the longest of all constellations in north–south dimension, Eridanus, the River. And then there is what the science of recent decades has revealed about Achernar as a sun. This is a star that stands out astrophysically as amazing and even unique in several ways among the brightest stars.

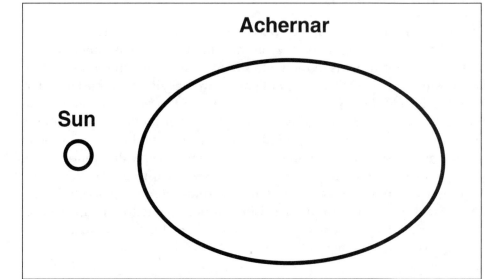

Relative sizes and shapes of the Sun and Achernar.

The River's End

Achernar was apparently not mentioned by Ptolemy, even though he should have been able to see it from the latitude of Alexandria (31° N). R. H. Allen remarks that this is one of several pieces of evidence suggesting that Ptolemy's catalog was not based on his own observations but on the now lost one of Hipparchus. Hipparchus, 5° farther north at Rhodes, would not have been able to see Achernar.

Achernar means "end of the river" and was originally given to the 3rd-magnitude star Theta Eridani, which was the farthest south part of Eridanus easily visible to the Arabs, who invented the name. Not until Renaissance times was "Achernar" transferred to the mighty 1st-magnitude star that lies 17° farther south than Theta Eridani (Theta eventually came to be called Acamar—just a different version of the same title).

The river of Eridanus begins with Beta Eridani, better known as Cursa, the "footstool" of Orion near his bright foot Rigel. From Cursa at –5° declination, the constellation meanders south and west (but mostly south) until it reaches Achernar at –57°. In all that length, there are no stars other than Achernar that are brighter than 3rd magnitude (though at least nine of the stars of Eridanus possess fairly well-established proper names: Cursa, Keid, Beid, Zaurac, Rana, Azha, Angetenar, Acamar, and Achernar). At least a few stars from neighboring constellations provide Achernar with some additional modestly bright company. In particular, magnitude 2.9 Alpha Hydri (second brightest star of Hydrus, the Male Water-Snake) is located just 5° from Achernar.

Lonely Achernar

I say that Achernar is lonely but as recently as 2000 Achernar and Fomalhaut lost the title of being the 1st-magnitude stars farthest in angular distance from any other 1st-magnitude star (each other). The new champion is Antares. The distance between Achernar and Fomalhaut is now 39°, 06'; the distance between Antares and Alpha Centauri is 39°, 07'.

Yet no one would ever call Antares lonely. It's not just how far away other 1st-magnitude stars are that counts in this consideration. Antares resides in the middle of the constellation that possesses the greatest number of stars brighter than magnitude 3.0, with several brighter than 2.5 or even 2.0—and they are all arranged along the magnificent curl of striking Scorpius. The next-door neighbor constellation is Sagittarius, which features the prominent Teapot pattern and its 2nd- and 3rd-magnitude stars. And the numbers of dimmer naked-eye stars—and Milky Way clouds—in and near Antares makes this region of the heavens gloriously crowded in dark skies.

By comparison, Achernar lies in a generally star-poor region and, as we saw, has few stars even as bright as 3rd magnitude anywhere near it. By the way, Achernar lies 61° from Alpha Centauri and 89° from Antares.

On the other hand, Achernar has an interesting visual connection with Canopus and with two of the great deep-sky wonders of the far-south heavens. The distance between Canopus and Achernar is 39° 26'—less than half a degree more than the Achernar–Fomalhaut separation. What's more, the three stars lie on nearly a straight line. Of course, an observer has to be located far enough south to have all three of these stars visible at once and appreciate the arrangement at the times when it can be seen. But if you observe from anywhere in the Southern Hemisphere at any time (as long as your sky is dark and clear), you can perceive a smaller formation—a wonderful triangle. The points of it are Achernar and the two greatest satellite galaxies of the Milky Way, the Magellanic Clouds. Achernar is only about 16° from the Small Magellanic Cloud (SMC) and 26° from the Large Magellanic Cloud (LMC). The remaining side of the triangle, the distance between the two clouds, is about 20°.

By the way, Achernar is lonely in an additional way: as far as we know, it has no companion stars.

The Flattest Star

Astronomers have used interferometry to discover that a number of our very brightest stars are pronouncedly oblate—that is, flattened at the poles, wider in equatorial diameter than in polar diameter. Altair, Vega, and Regulus are all known to be outstanding cases, and the cause can be attributed to the rapid rotation of these stars. But Achernar is easily the most extreme example of spin-induced flattening known among the very bright stars.

A study in 2003 found that Achernar's equatorial diameter was an astonishing 56 percent greater than its polar diameter. The latest interferometric data put the equatorial diameter at 11.8 times that of the Sun and the polar diameter at only 7.6 times that of the Sun. The equatorial spin velocity of Achernar must be at least 250 km/sec. That's fast enough for the star to be losing a lot of its gas—thousands of times faster than our Sun does.

Achernar is, in fact, an example of a Be or B-emission star. Such a star is of B spectral class—Achernar is B3— and has a circumstellar disk of gas that adds emission lines to its spectrum. The first of these stars identified was Gamma Cassiopeiae (the middle star in the W or M of Cassiopeia). Another bright example is Delta Scorpii. Both of these Be stars eventually engaged in dramatic brightenings and brightness variations. Delta Scorpii had always been measured as a magnitude 2.3 star until the opening years of the twentieth century. In recent years, the star has tended to be considerably brighter, fluctuating to as bright as magnitude 1.6 a few times. Might Achernar someday enter into a state of marked variability? That certainly seems possible. If it behaved the way Gamma Cassiopeiae and Delta Scorpii did, it would perhaps start rivaling Alpha Centauri in brightness for a while. But there's no telling whether Achernar will do so this year or not until some time millennia in the future.

Achernar also belongs to the class of "Lambda Eridiani" stars that show very slight but regular variations. These variations are believed to be caused by either pulsations of the star or dark starspots being pointed in our direction each time the star rotates around.

More on Achernar as a Sun

Achernar is located about 144 light-years from Earth (a little less than twice the distance of Regulus). It is believed to be only a few hundred million years old. It is still fusing hydrogen into helium. Its surface temperature is uncertain, but is most likely to be around 14,500 K—quite hot, though it could be even hotter. Depending on how hot Achernar really is, it could be emitting anywhere from 2,900 to 5,400 times as much energy as our Sun (much in the ultraviolet range).

Achernar's mass is probably about six to eight times the solar value. But if the star continues to lose mass rapidly from its equator, it will probably never become a supernova. Its eventual fate might be to turn into a massive white dwarf like Sirius B.

This distinctly blue-white star has two neighboring star systems within about 5 light-years of it. One is a Sun-like star about 3.5 light-years away. The other is an orange main-sequence star 4.9 light-years from Achernar.

17 BETELGEUSE

Betelgeuse was given the Greek letter designation Alpha Orionis despite the fact that Rigel—Beta Orionis—is a magnitude 0.1 star. But perhaps that is not unfitting considering that Betelgeuse has so many amazing distinctions, arguably as many as any star known.

Betelgeuse is the classic red giant, the most famous. It is the brightest "red"—well, deep orange-gold or spectral type M—star in the Earth's sky. It may also be the reddest, biggest, and coolest of all the very bright stars (though Antares rivals it in these categories). It is also the 1st-magnitude star with the greatest range of brightness variation. Betelgeuse is the most visually outstanding star of the most visually outstanding of all constellations. And then there is its name. Betelgeuse has the most distinctive, indeed the most conversation-provoking, of all star names. The name has even, in comically altered form, provided the title for a popular Hollywood film.

The Name and Lore of Betelgeuse

Let's start with that name of the star. Paul Kunitzsch provides the most thorough and authoritative discussion of it. He states that the original, early Arabic name for Betelgeuse was *yad al-jauza*, which means "the Hand of *al-jauza*," and adds, "This *al-jauza* has been translated as 'the Giant' and 'the Middle One.'" Kunitzsch notes that it was the early Arabic name for Orion, regarded as a feminine figure. "The root *jwz* can mean 'middle' and the word *al-jauza* is structured as a feminine adjective, thus *al-jauza* may mean 'the female one, having something about her related to the middle.'" But the middle of what is unknown. The best theory may be "the middle of the sky"—that is to say, the celestial equator, upon which Orion lies. Another theory is that Orion (or part of Orion) was being compared to a special kind of sheep that had a central belt or spot of white on it. But, as Robert Burnham Jr. says, most of us will prefer to think of Orion as representing a more gallant and majestic figure than that of a sheep!

But the real problems began when *yad al-jauza* was transliterated into Latin. Kuntizsch explains: "The first medieval transliteration into Latin was

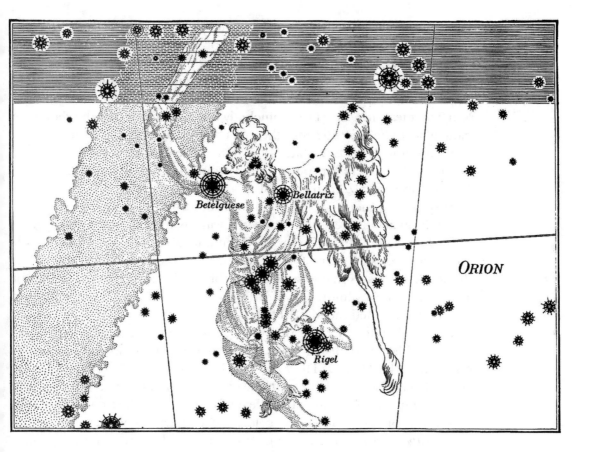

bedalgeuze, mistaking the initial Arabic word '*bat*' [properly *ibt*], for 'the Armpit' of *al-jauza*, giving rise to the corruption 'Betelgeuse' seen today." Thus by Renaissance times both the '*y*' and the '*d*' of the first part" of the early Arab name had been corrupted.

In the nineteenth century and into the twentieth, the name was still sometimes written as Betelgeuze or Betelgeux. But in recent decades, Betelgeuse has become the standard spelling. The big question is: How do you pronounce it?

Most astronomers and scholars of star names surely prefer to say BET-el-jooz or BET-el-joos. But many people seriously believe that the accepted pronunciation is BEET-el-joos. Others don't care which form is preferred. They are just amused to call this big, famous, oddly named star "beetle juice." *Beetlejuice* is the title of a movie directed by Tim Burton and whose title character is a comically cantankerous ghoul brought to life (or, technically speaking, death) by actor Michael Keaton. Do the movie and character take their name from that of the red giant star? Absolutely. *Sky & Telescope* magazine published a Letter to the Editor from the movie's script writer, Malcolm McDowell, who explained, "During the four years I spent with the project in Hollywood, I was repeatedly delighted—and somewhat astonished—by people who responded to the title *Beetlejuice* with the question, 'Oh, you mean like the star?' Somebody even suggested that the sequel be named *Sanduleak −69 202* after the precursor of Supernova 1987A."

By the way, Kunitzsch states how Betelgeuse would be pronounced in medieval Arabic. The pronunciation would be BET-ul-jow-ZAY—with a stronger accent on ZAY than on BET.

Surprisingly, there doesn't seem to be a lot of lore connected to Betelgeuse. Probably the legends connected with its majestic constellation as a whole have stolen the place for Betelgeuse in particular. We do know that the ancient Sanskrit name for the star was Bahu, which means "the Arm." But R. H. Allen says that what was probably really meant was the foreleg of an Antelope being hunted by Sirius—the hunter Mrigavyadha. Betelgeuse has also been called the Martial Star.

Burnham notes that the twentieth-century horror fantasy writer H. P. Lovecraft identified Betelgeuse as the home of the infinitely wise beings called the "Elder Gods." The Elves of J. R. R. Tolkien's Middle-earth called either Betelgeuse or Aldebaran by the name Borgil, "Fire-Star." Robert Burnham Jr. himself came up with a beautiful name and lore-connection for Betelgeuse. Speaking about the color of Betelgeuse, he wrote, "The exotic word *padparadaschah*, used in India to designate the rare orange sapphire, might be an appropriate name for Betelgeuse."

The Brightest Red Giant

For many decades, Betelgeuse was regarded as being, among the ranks of the 1st-magnitude stars, the largest and reddest and coolest of stars and most luminous of the red giants. Betelgeuse, winter's great red giant, was regarded as surpassing in these attributes its only competition—Antares, summer's great red giant. In the past decade or so, however, new estimates of the distances of these two stars has called into question the old claims of Betelgeusian superiority. If Betelgeuse is as close as 425 light-years and Antares as far as 600 light-years, Antares would be the more luminous star.

Our uncertainty about these distances doesn't change one fact, however: Betelgeuse is unquestionably the red giant of greatest apparent brightness in our sky. In this book, I am following the authorities that place the average brightness of Betelgeuse at magnitude +0.5—which makes it the tenth-brightest star in all the heavens, seventh brightest visible at midnorthern latitudes. Some sources place the star's average apparent magnitude a little brighter— +0.4—and others put it dimmer— +0.7.

All such figures make it brighter than Antares, which apparently never gets brighter than +0.9 and almost never varies outside the range of 0.9 to 1.1. The common range of Betelgeuse might be from about 0.3 to 0.8. But it seems to venture outside this range more often than Antares goes outside of its own. The extreme range of Betelgeuse is also greater and, unlike Antares, features a bright end. Antares has an extreme range from 0.9 to 1.8. Betelgeuse has an extreme range from about –0.1 to +1.3—but I have read of observations that placed Betelgeuse as dim as +1.5 or +1.6 . At its rare brightest, Betelgeuse outshines Rigel. At its rare dimmest, it barely exceeds the brightest stars of Orion's Belt.

Perhaps the most valuable and exciting record of the brightness of Betelgeuse is that obtained by a particular skilled observer over a long period of time. Let's look at what the estimates of two such observers show.

The Best Brightness of Betelgeuse

The variability of Betelgeuse was never more exciting than in certain periods in the mid-nineteenth century. During those spells, the star was carefully monitored by the great Sir John Herschel. Surprisingly, he may have been the first person to notice these changes. They were dramatic in 1836–1840 but were even more impressive in 1849–1852. In December 1852, Herschel wrote that he thought Betelgeuse was then "actually the largest star in the northern hemisphere." By this he meant that Betelgeuse appeared brighter than any star in the north celestial hemisphere—thus brighter than Capella and perhaps Arcturus. Herschel's statement would suggest a magnitude of perhaps –0.1. Betelgeuse may have been roughly as bright again in 1894. Robert Burnham Jr., writing in the 1970s, said that the star had particularly high maxima of brightness in 1925, 1930, 1933, 1942, and 1947, "while during the decade 1957–1967 only slight and uncertain variations were detected." AAVSO graphs of brightness estimates show that Betelgeuse probably reached magnitude 0.2 in 1933 and 1942, and faded to dimmer than 1.2 in 1927 and 1942. But the most interesting record of Betelgeuse's brightness in the twentieth century that I know of is that of astronomer and former *Sky & Telescope* editor in chief Joseph Ashbrook.

Ashbrook studied Betelgeuse between 1937 and 1975 and in that period estimated its extreme values of brightness as being –0.1 and +1.1. Ashbrook's record shows that the variations of Betelgeuse are usually small and gradual but not always: in February 1957, for instance, he saw the star brighten by 0.4 magnitude and dim back by 0.4 magnitude all in the course of just two weeks.

In the past thirty years or so, my brightness estimates of Betelgeuse have not been regular and have usually been casual. But I know that I've seen it at least as dim as Aldebaran (magnitude +0.87) on one occasion and, just a few years ago, saw it at least as bright as magnitude 0.3, perhaps a bit brighter.

The brightness of Betelgeuse definitely bears watching. No other 1st-magnitude star is markedly variable in brightness and reasonably often. As with other "semiregular" variable stars, Betelgeuse has unpredictable brightness fluctuations superimposed on several approximately regular periods of brightness variation. In the case of Betelgeuse, the main period seemed to be about 5.7 years, with shorter superimposed periods that vary between 150 and 300 days.

How Big Is Betelgeuse?

The brightness variations of Betelgeuse are interesting enough but they are accompanied by changes in the star's size. These pulsations in the size of red giants have been compared to the beat of a vast heart and to the breathing of a vast beast. An AAVSO publication discusses the probable cause of the pulsations:

> Astronomers think the outer layers of the star expand slowly for several years and then shrink again, so the surface area alternately increases and decreases, and the temperature rises and falls, making the star brighten and dim. Red supergiants pulsate this way because their atmospheres are not quite stable. When the star is smallest, the atmosphere absorbs a bit too much of the energy passing through it, so the atmosphere heats and expands. As it expands, it becomes thinner. Energy then passes through the outer layers more easily so the gases cool, and the star shrinks again.

Betelgeuse's width may vary by as much as 60 percent during this cycle. So one must speak of an average diameter.

Betelgeuse was the first star to have its angular diameter measured. This was way back in 1920, when the measurement was accomplished with the 100-inch telescope at Mount Wilson, using a "beam interferometer" invented by Albert A. Michelson, the first American to win a Nobel Prize in

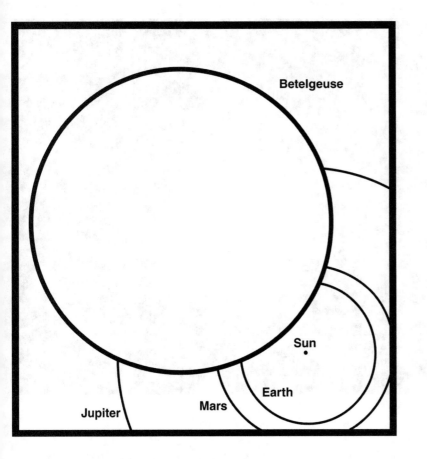

Size of Betelgeuse at possible largest in pulsation cycle compared to orbits of planets in our solar system.

physics (1907). This experiment found that the average angular diameter of Betelgeuse was about 0.044".

In 1975, Betelgeuse became the first star other than the Sun to have brightness features on its surface identified. Researchers used "speckle interferometry," image intensifiers, and image processing to reveal a mottling of the disk with dusky areas, presumed to be the Betelgeusian equivalent of sunspots.

In 1995, Betelgeuse became the first star to have its disk directly imaged by a telescope. The Hubble Space Telescope imaged the star's atmosphere in ultraviolet and revealed a bright spot on the giant star's surface. This spot was more than ten times the diameter of Earth (about Jupiter-size) and had a temperature at least 2,000 K hotter than the overall surface of the star.

How large does Betelgeuse appear in Hubble imaging? The star looks to be well over a thousand times wider than our Sun, roughly a billion miles—a sphere large enough to easily contain the orbit of the planet Jupiter. But is

Hubble Space Telescope direct image of Betelgeuse's globe. The light area is an actual hotter region.

this the size of the surface of the star or the atmosphere in ultraviolet out to as far as Hubble could record? The latter would seem to be the case; more recent sources quote an average diameter of "only" about 650 times that of the Sun. This would make Betelgeuse average about 3 AU—large enough to just contain the orbit of Mars, but a little smaller than Antares.

In truth, even the outer visible layers of these red supergiants are more tenuous than the best vacuum we could create on Earth. In any case, we probably don't know enough about red supergiants to understand the various outer layers. How should we define where the star ends and its atmosphere begins? In the case of the Sun, it's pretty easy to make the distinction between the blindingly bright photosphere and the chromosphere and corona that lie above (outside of) it. Like Antares, Betelgeuse seems to be surrounded outermost by a nebula, up to several light-years across.

By the way, Betelgeuse is at least larger in angular size than Antares. Or than any other star except for R Doradus—a red giant that is smaller in true size than Betelgeuse but much closer to us.

Star within a Star?

Now we come to what might be the most amazing possibility of all about Betelgeuse: the possibility it has a companion star that actually orbits *within* it.

This was a claim made in 1985 by a researcher who based it on speckle interferometry of Betelgeuse. The data suggested the presence of two companions of Betelgeuse. One was a star that orbits about 40 or 50 AU from Betelgeuse. But the other was a star whose mean distance from the center of Betelgeuse was 5 AU. If such a star exists, it's possible that it would actually spend at least part of its orbit passing under the surface of Betelgeuse and into what we would truly call the star, not just its atmosphere.

Such a situation would be possible because of the extreme tenuousness of the outer layers of Betelgeuse. In a related story, a 2006 study discusses a "brown dwarf" star that survived within the outer layers of a red giant until the latter puffed those layers off in its dying phases and became a white dwarf. So perhaps a star really could survive a long time orbiting within Betelgeuse before it was slowed enough to spiral down into a collision with the denser central regions of the behemoth star.

Unfortunately, no later study I'm aware of has substantiated the 1985 claim that Betelgeuse has companions—including (literally including!) one very close to it.

When Will Betelgeuse Blow?

Red giants release most of their energy as infrared radiation, undetectable to human eyes. Betelgeuse is an M2 star (of luminosity class Iab), whereas Antares is M1.5. This would seem to suggest that Betelgeuse has a cooler surface. But experts say that there is probably not a strong correlation between surface temperature and spectral subtype among different M-type stars. The energy released by Betelgeuse is estimated to be only about 13 percent visible light. Depending on what percentage of Antares's output of electromagnetic radiation is visible, could it be that Betelgeuse produces more total radiation than Antares? The claim has long been made that if we were able to record all wavelengths of light with our eyes, Betelgeuse would appear brighter than any other star in the night sky—perhaps as bright as Venus now does.

But how bright will Betelgeuse get in visible light in our sky if it goes supernova?

Betelgeuse should go supernova sometime between now and a few million years hence. If it goes off early in this period, it should be close enough to us to burn brighter than the Full Moon in our sky. But perhaps we shouldn't hope for this show in our lifetime. For it also may damage and disturb Earth's biosphere at least slightly with a powerful dose of hard radiation.

Betelgeuse is the classic "red giant" star—a stage in stellar life that perhaps every star we see with the naked eye in our sky is either going through or will go through someday. Our Sun will probably never swell to more than about the diameter of the Earth's orbit and probably not for something like 5 billion years (though recent studies suggest that the Earth could become unlivable from the earlier phase of a swelling red Sun as "soon" as about 1 billion years from now). Our Sun is not massive enough to ever go supernova. But Betelgeuse is thought to have started its life with maybe as much as thirteen to seventeen times more mass than the Sun. The behemoth is unlikely to ever lose enough of its mass to stellar wind for it to avoid eventually going supernova.

18 ▶ BETA CENTAURI

As recently as 2004, astronomers were amazed to learn that Beta Centauri was less than two-thirds the distance from Earth than had previously been thought. That still left Beta Centauri about seventy-seven times farther than its famed close neighbor in the sky, Alpha Centauri. It still kept it an impressively luminous sun—or rather twin suns, with a third luminous star not far away. It also didn't change the fact that Beta Centauri is in competition with Alpha Crucis, Beta Crucis, and Spica for title of bluest and hottest of the 1st-magnitude stars. But, among other things, the new distance placed Beta Centauri in the same neighborhood of space as Alpha, Beta, and Delta Crucis, the three blue bright stars of the Southern Cross.

The Three-Named Partner of Alpha Centauri

Before we explore the spectacular physical nature of the Beta Centauri system, let's step back and ponder the appearance and names of this star.

Alpha and Beta Centauri lie just less than 4½° apart in the sky. As we discussed earlier in this book, that makes it the second tightest naked-eye pair of very bright stars in the heavens after Alpha and Beta Crucis—until about 2166, when Alpha and Beta Centauri will become the tightest. But Alpha and Beta Centauri are a much brighter pair than the two brilliant Crux stars. Alpha Centauri shines almost a magnitude brighter than Beta Centauri in our sky but that means Beta Centauri burns at magnitude 0.61 (with dimmings to 0.66 in a period of 0.16 days). Such brightness makes it usually the eleventh-brightest star of the night, slightly exceeding Alpha Crucis and greatly exceeding Beta Crucis.

Our discussion of the lore of Alpha Centauri included a few titles and picturings the star has shared with Beta Centauri. (One title for Beta Centauri alone, however, is the Chinese Mah Fuh, which R. H. Allen mentions and says means "the Horse's Belly.")

Additional lore is inherent in a few of the names for Beta Centauri that have been adopted for it in modern times. The Greek-letter designation "Beta Centauri" doubles as the most common name for the star. But other names frequently used for this star include Hadar and Agena.

Agena has been applied to Beta Centauri in modern times (perhaps first by Burritt in the nineteenth century). Although R. H. Allen didn't recognize the meaning, other scholars have figured that Agena is a combination of the Greek *alpha* and the Latin *genu*, which means "knee" (note the English word for "bending the knee," typically in an attitude of worship—*genuflect*). The "knee" part of the name Agena makes sense because Ptolemy mentioned Beta Centauri as marking the knee of the Centaur's left front leg. But the initial letter A, if for alpha, is incorrect for Beta Centauri. Strangely, a similar mistake was made in giving the name "Bungula" to Alpha Centauri. Here, the B would seem to be for *beta*, and *ungula* is Latin for "hoof"—Ptolemy placed Alpha Centauri on a foot of the Centaur.

More popular than Agena for Beta Centauri has been Hadar. According to Paul Kunitzsch, the early Arabs invented the names *hadari* and *al-wazn* for a pair of stars but then the "scientific" Arabs of the later Middle Ages decided to identify these names with either Alpha/Beta Centauri or Alpha/Beta Leporis (stars in Lepus, the Hare). Much more recently and, says Kuntizsch, "arbitrarily," one of the names—Hadar—was applied to Beta Centauri.

Kunitzsch says that *hadari* is an untranslated proper name of unknown significance. R. H. Allen translates it as "ground" (the idea of a bright star seen very near the horizon—Beta Centauri was low in the south for Arab observers). On the other hand, Guy Ottewell, in his *Astronomical Companion*, derives Hadar from the Arabic *al-Hadar*, meaning "the settled land" ("as opposed," writes Ottewell, "to the desert").

After the naked-eye view of Beta Centauri and the star's lore, the next thing to consider is its telescopic appearance. Earlier in the book, we saw that Alpha Centauri is a glorious and usually very easy (widely separated) double star through a telescope. Beta Centauri is also a visual double but is a very challenging one. The companion, which we could call Hadar B, shines at magnitude 3.9 but is only 1" from the primary. The companion takes about 250 years to orbit the primary, with a minimum distance of 110 AU. It shines with a luminosity several hundred times that of the Sun and is a quite hot star of spectral class B8.

The Twin Giants of Hadar A

The great interest in the Beta Centauri system, however, is the primary—especially because in recent years interferometric studies have revealed that this primary is really itself a close binary star.

Hadar A consists of nearly identical stars that orbit around each other in a period of 357 days. Their average separation is 2.59 AU but the orbit has high eccentricity that takes the stars as close as 0.46 AU (almost as close together as the Sun and Mercury) and as far as 4.72 AU (not all that much less than the gulf between the Sun and Jupiter).

The 2004 study mentioned at the start of this chapter suggested that the stars really were just about twins. Each was calculated to weigh in at 9.1 solar masses. One of the stars had an absolute visual magnitude of −3.85, the other −3.70. The diameter of each may be about eight times that of the Sun. One of the twins is—or perhaps both are—a Beta Cephei type variable, brightness altering very slightly in a period of less than 4 hours. Both stars are luminosity class III objects—giants—and probably B1 spectral type. Their exact surface temperatures might be as high as 25,000 K or as low as 22,500 K (more like that of a B2 star). So they may be almost as hot as the major components of Alpha Crucis and Beta Crucis.

But the 2004 study was not the last word. In 2006, a different group of researchers came up with a distance for Beta Centauri larger than 330 light-years but still very much smaller than the 530 light-years derived from the Hipparcos satellite parallax measurements in the late 1990s. The new figure was 352 ± 13 light-years. This is a bit farther from Earth than Alpha Crucis is but still, within the margin of error, about as far from us as Beta Crucis and Delta Crucis are. The new study finds the components of Hadar A a little larger—10.7 ± 0.1 solar masses and 10.3 ± 0.1 solar masses. It also comes up with a new, younger age for Beta Centauri—just 14.1 ± 0.6 million years old.

What does the future hold for the twin giants of Hadar A? Their cores are about to use the last of their hydrogen for fusion. The two will be swelling much more and becoming red giants. They will interact in major ways. One or both of them could eventually become supernovae. Or they could both become massive white dwarfs.

What do the twin suns of Hadar A currently look like from the vicinity of Hadar B? They would appear an average of a bit more than 1° apart and appear as tiny blazing disks only about 2 arc minutes wide.

ALPHA CRUCIS

19

A lpha Crucis is the brightest star of the Southern Cross. It marks the foot of the famous pattern, its southernmost star, the one which is at the bottom of the directly upright cross when the cross stands at its highest in the due south. The Southern Cross as a whole is so much more famous and attention grabbing than even its brightest individual star, that Alpha Crucis has virtually no lore of its own. Likewise, it has no proper name better than Acrux (pronounced *AY-cruhks*)—which is nothing but a contraction of the Greek-letter name or designation. This title for the star was apparently originated by Burritt in the nineteenth century.

As we'll see, when viewed through a telescope, Alpha Crucis turns out to be one of the most splendid of all bright double stars—a less brilliant pair than Alpha Centauri but a more brilliant pair than Castor A and B. Yet much like other blazing stars of B spectral type in this region of the heavens—for instance, Beta Centauri and Beta Crucis—Alpha Crucis is really a more complicated system of multiple stars, fascinating to detail.

Acrux as Visual Double or Triple

Alpha Crucis shines at a total apparent magnitude of 0.77. That's almost identical to the brightness of Altair, and some lists place Altair just ahead of Acrux. In our tally here, however, Acrux is the twelfth-brightest and Altair the thirteenth-brightest star in the nighttime heavens.

We've already described the naked-eye scene of grandeur in this part of
the sky: how Alpha and Beta Crucis are (for another century and a half, at
least) even closer together than Alpha and Beta Centauri; how a line drawn
from Beta Crucis through Alpha Crucis points toward the south celestial
pole; how the Southern Cross stands in the gap between those brightest
gems of Centaurus and major marvels of Carina (the Eta Carinae Nebula
and more). But what does Alpha Crucis itself look like in a telescope?

The view through the eyepiece shows us a magnitude 1.25 star just 4"
from a magnitude 1.64 star (James Kaler gives 1.33 and 1.73 as the respective
magnitudes of these objects, Alpha-1 and Alpha-2 Crucis). According to R.
H. Allen, this doubleness of Acrux was first discovered by "some Jesuit missionaries sent by King Louis XIV to Siam in 1685." But these two stars are not
all there is to the telescopic sight. A generous 90" south–southwest of the
brilliant pair—far enough out to spot in binoculars—shines a 4.9-magnitude
third star. It is a B-type. This third star *seems* to share the same motion
through space as the Alpha-1/Alpha-2 pair, and has been named Alpha Crucis C. But it is likely to be more than twice as far from Earth as the other two
stars, merely lying along the same line of sight.

The Real Acrux Trio

Alpha Crucis C may not be a member of the same star system as the Alpha-1 and Alpha-2 stars. The system, however, is triple nonetheless: Alpha Crucis-1 is, itself, composed of two close-together suns.

Alpha-1 Crucis is a spectroscopic double. Its combined spectral type is a B0.5 and its luminosity class is IV (subgiant). This compares with Alpha Crucis-2's being a B1 star of luminosity class V (dwarf—that is, main sequence—star). The expected surface temperatures for Alpha-1 Crucis and Alpha-2 Crucis would be 28,000 K and 26,000 K, respectively. This would make them likely the hottest of all the 1st-magnitude stars. Given the system's estimated distance of 320 light-years from Earth, the luminosity of Alpha-2 Crucis would be 16,000 times that of the Sun and the luminosity of Alpha-1 25,000 times that of the Sun. But I repeat: Alpha Crucis-1 is a double. What do we know about its two components?

The two stars of Alpha-1 are calculated to be about fourteen and ten times as massive as the Sun. They are located about 1 AU from each other and they take only seventy-six days to orbit each other. Meanwhile, the single star of Alpha-2 Crucis weighs in at about 13 solar masses. It is roughly 430 AU from the tight Alpha-1 duo.

Alpha-2 Crucis and the brighter component of Alpha-1 Crucis are massive enough that they will probably someday go supernova. The fainter and less massive of the Alpha-1 duo may escape that explosive phase and become a massive white dwarf.

What would Alpha-1 look like from Alpha-2? According to James Kaler, it would appear as a naked-eye double star—a bit over one-tenth of a degree separating the two blue-white specks whose combined light would be about ten thousand times greater than that of a Full Moon on Earth. That would, of course, be dangerously bright to look at unless attenuated by a thick atmosphere, perhaps as the awesome object rose or set on some planet of Alpha-2 Crucis.

Acrux in the Scorpius-Centaurus Association

The most glorious clutch of youthful, prodigal suns in the direction of Centaurus and Crux is the one that includes the following stars (with their distances from Earth included): Alpha Crucis (320 light-years), Beta Crucis (350 light-years), Delta Crucis (360 light-years), and Beta Centauri (352 light-years). How glorious must these stars be as seen from each other? This far from Earth, Hipparcos parallaxes give a distance with an uncertainty of about 8 to 10 light-years. Even those of these stars that are about 30 light-years apart would appear like Venuses in each other's sky.

20 ⟩ ALTAIR

A ltair is distinctly the second-brightest star that most of us can see all the way from spring's Arcturus to winter's Capella. Altair is also simply beautiful in its positioning—in a wonderful set of different ways. It is positioned wonderfully between two stars; wonderfully in the very apt geometric pattern that represents Aquila, the Eagle; wonderfully at the southeast corner of the grand Summer Triangle. It is also wonderfully high (not neck-strainingly high) most of the summer and early fall, wonderfully at the same height as Vega as the two near the horizon, wonderfully poised on the opposite shore from Vega at the most majestically rifted part of the Milky Way.

The Eye of the Eagle

Altair comes from the Arabic *at-Ta'ir*, "the flying one." The Arabs called Altair and Vega, or Aquila and Lyra, *an-Nasr at-Ta'ir* and *an-Nasr al-Waqi*, "the flying (or soaring) eagle" and "the swooping (or stooping) eagle," although *an-Nasr* could also mean "vulture." The pattern of the constellation Lyra could be imagined as a bird with its wings drawn nearly all the way in to its body—as when stooping down to seize its prey. The pattern of Aquila can most easily be imagined as a bird with its wings widespread. It even looks as if Aquila may be soaring and slowly turning, wheeling around on the high winds.

The name Aquila is Latin for "eagle." And it's interesting that the imagining of an eagle here probably goes back to far before the Romans and Greeks. Even the perhaps somewhat conservative authority Paul Kunitzsch says that the name *an-Nasr at-Ta'ir* "has probable origins among the Babylonians and Sumerians, for whom Alpha Aql was 'the Eagle Star.'" Alpha Aquilae is Altair, of course (it greatly outshines any other star in Aquila). R. H. Allen, writing more than a hundred years ago, says that Aquila is "supposed to be represented by the bird figured on a Euphratean uranographic stone of about 1200 B.C." and known "on the tablets" by a title that means "the Eagle, the Living Eye."

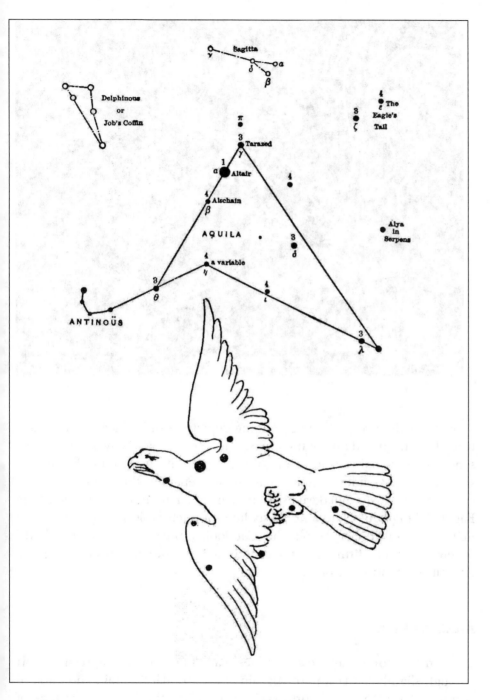

Aquila and Altair.

Aquila and
Altair.

Is Altair the Eye of the Eagle? Some depictions of it place it at the juncture of the head and one of the wings. But, to me, it has always made perfect sense to see it as the head or eye of the eagle, with its beautifully flanking stars Beta and Gamma Aquilae marking like epaulets the shoulders of the eagle with its outstretched wings. Tennyson has a famous poem entitled "The Eagle: A Fragment," and a noble six lines of poetry it is. But it is elsewhere he writes: "Only the eagle, only he/Can look at the sun/And not go blind." Surely it would be fitting for noble Altair to mark the mighty eye of the legendarily most noble of birds?

Altair in China

Titles in lore for Altair are sometimes really titles for the impressive short (though slightly bent) line of Alshain (Beta Aquilae), Altair, and Tarazed (Gamma Aquilae). In Hindu skylore, they are the Three Footsteps of Vishnu, Vishnu being the third in the triad of Brahma (Creator of Worlds), Shiva (Destroyer of Worlds), and Vishnu (Preserver of Worlds). The three stars have also been known as the Family of Aquila and the Shaft of Aquila.

It can also be either Altair alone or this line of three stars that represents the Cowherd in the greatest of all Chinese star-myths, the one of the Weaving Princess and the Cowherd.

I relate that story briefly in the chapter on Vega (Vega, or Vega and its little constellation Lyra, is the Weaving Maiden). But another hauntingly beautiful Chinese story is that of the voyage of Chang K'ien up the Yellow River and into the Milky Way. Gertrude and James Jobes tell wonderfully how this voyager awoke to find himself sailing through a land of blue and silver light where the stars looked nearer and brighter than he had ever seen them and were reflected by the thousands in the river. Chang K'ien saw a herdsman on one bank of the enchanted river and a maiden weaving on the other bank. He stopped to ask her what land this was but she would only tell him to take the shuttle from her loom and, when he returned home, to show it to Kun P'ing the court astronomer. When Chang K'ien eventually got home, he followed the maiden's instruction. The wizardly astronomer told him that it was the Weaving Princess and Cowherd he had seen and met, that he had been sailing in the heavens on the Milky Way. The astronomer had observed a "visitor star"—a comet or a nova?—between the Princess and Cowherd, and that star must have been Chang K'ien.

This story is supposed to have taken place during the Han Dynasty over two thousand years ago, and I've always wondered if it was derived from an observation of an actual nova in Aquila that occurred around that time. The most brilliant nova of modern times occurred in western Aquila north of the Scutum Star Cloud in 1918. When first noticed, it was already a 1st-magnitude object brighter than Altair. On June 9, Nova Aquilae 1918 reached a stunning peak brilliance estimated as magnitude –1.4—virtually the equal of Sirius—before fading away.

Noble Positionings of Altair

As I said at the start of this chapter, Altair's various positionings—in small asterism, constellation, large (transconstellational) asterism, ascendancy of itself, and setting with Vega—all happen to be stately and grand.

The compact though slightly crooked line formed by Alshain, Altair, and Tarazed is beautifully distinctive. A typical pair of binoculars, with a field of view about 5° wide, can fit in the entire asterism. A straight line from Tarazed to Alshain is 4° 46' long. Tarazed is 2° 4' from Altair. Alshain is 2° 42' from Altair. Alshain shines at magnitude 3.7; Altair, at 0.77; and Tarazed, at 2.7 (thus Altair is almost exactly 2 magnitudes brighter than Tarazed and 3 magnitudes brighter than Alshain). Alshain is about 45 light-years from Earth; Altair, 16.7 light-years; and Tarazed, 500 light-years. The true brightness

The Cowherd (Altair) and the Weaving Maiden (Vega) cross the Milky Way on the bridge of magpies.

of Alshain is a little less than that of Altair, but rather distant Tarazed is roughly one hundred times more luminous than Altair. Using binoculars or a small telescope, the colors of the three stars can be noticed. The B-V color index for each is 0.2 (Altair), 0.9 (Alshain), and 1.5 (Tarazed). Those values translate as white with perhaps a very slight tinge of yellow (Altair), rich yellow (Alshain), and distinctly orange-gold (Tarazed).

Altair is also part of a trio of stars that forms not a small bent line but a huge triangle—the Summer Triangle. The stars of the Summer Triangle are blue-white Vega (magnitude 0.03) in Lyra, ever-so-slightly yellow-white Altair (magnitude 0.77) in Aquila, and pure white Deneb (magnitude 1.25) in Cygnus. Thus Altair is about ¾ magnitude dimmer than Vega, and Deneb is about ½ magnitude dimmer than Altair. The angular distances between them are 34° from Vega to Altair, 38° from Altair to Deneb, and 24° from Deneb to Vega. Their true distances from us in space are 16.7 light-years (Altair), 25 light-years (Vega), and Deneb at somewhere between about 1,500 and 2,600 light-years away. If Deneb is at the latter distance, its luminosity is an astonishing 160,000 times that of the Sun. But the relatively near Altair and Vega are respectively 10.6 and 37 times as radiant as the Sun.

The orientation of Altair and the Summer Triangle with respect to the horizon and sky varies beautifully during the year and the night. At latitude 40°N, Deneb rises more than one hour after Vega but Altair not until almost four hours after Vega. Altair reaches its culmination at 11:00 P.M. daylight saving time on August 20, more than an hour after Vega and almost an hour before Deneb. For those observers at 40°N, Altair and Vega are at about the same altitude in the west by the time that they are about 45° high. They continue at similar altitude much of the way down the sky but then Altair gets lower faster and sets about an hour before Vega (more northerly and easterly Deneb doesn't set until about three hours after Vega).

As seen from around 40°N, Altair is about 45° high when it is in the southeast and 45° high when it is in the southwest. Halfway through the four hours between those two positions, Altair culminates about 60° above the south horizon. During those four hours, it can be observed nobly high with a slight tilt up of your head. It is, in this period, a grand sight which can easily be enjoyed separately from Vega and Deneb because those two stars are so high the observer must really crane his or her neck to spot them.

Altair as a Sun

Altair's spectral type is A7 and its luminosity class is V—a main sequence star. It is the coolest of the three Summer Triangle stars, with a surface temperature

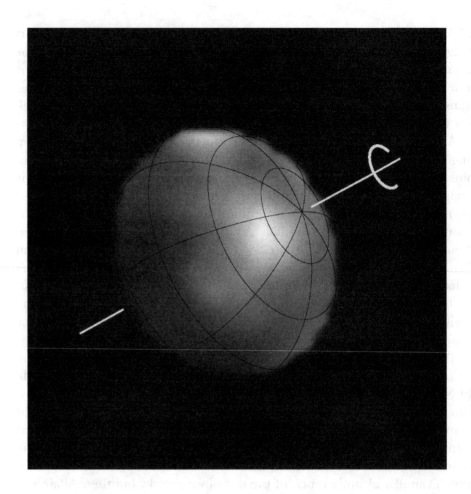

Altair's globe, imaged by CHARA interferometer.

of about 7,550 K. Altair's mass and its average diameter are both about 1.7 or 1.8 times that of our Sun. But notice the phrase "average diameter." Altair is decidedly oblate, its equatorial diameter about 14 percent greater than its polar diameter. Altair was the first of the bright stars found to have very fast rotation (by observation of the widening of its spetral lines). Its rotational speed at its equator is at least 219 km/sec (maybe greater depending on the tilt of Altair's axis). That's over one hundred times faster than the rotation of our Sun at its equator. The rotational period must at most be about ten hours but may be considerably less.

Altair lies at a distance of 16.7 light-years, making it summer's closest star plainly visible to the naked eye. Its distance is about 1 million times greater than the distance to our Sun. Altair is about: 4 times farther from us than Alpha Centauri, twice as far as Sirius, 1½ times as far as Procyon, ⅔ as far as Vega, ½ as far as Pollux, and ⅓ as far as Castor.

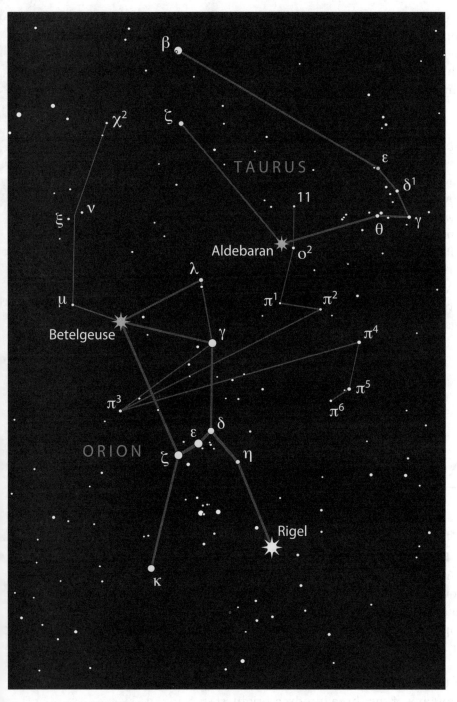

Changed view of Orion and Taurus as seen from Altair.

Altair is moving relatively fast though space from our viewpoint. Its proper motion is changing its position by about 1° each 5,000 years. The star is very slightly variable in brightness. It is classified as a Delta Scuti variable with nine different periods of variation that range from 50 minutes to 9 hours.

The View from Altair

Imagined planets circling Altair have been the setting for a number of well-known science-fiction stories. None is as famous as the 1950s movie classic *Forbidden Planet*, with its theme based on Shakespeare's *The Tempest* and its iconic robot.

Procyon and Beta Aquilae are both 28.0 light-years from Altair. Sirius is 25.1 light-years from Altair, and Procyon is 21.9.

A lovely sight in Altair's sky would be Vega. At a distance of 14.8 light-years from Altair, it would shine at about magnitude –1.1. As seen from Vega, Altair itself would shine at about –0.3.

21 ▶ ALDEBARAN

The star which literally marks the Bull's Eye—the bright eye of Taurus, the Bull—cannot help but be itself a bull's eye to our attention.

Aldebaran has remarkable distinctions of its own, some of them little known. Among the brightest stars, it is the deepest orange of the K stars, though not as ruddy as the M stars such as Betelgeuse. One thing that most people don't know is that Aldebaran was once Earth's brightest nighttime star and, when at peak, was our brightest star of any in almost the past million years. Another thing few people realize is that Aldebaran is now receding from us far more swiftly than any other 1st-magnitude star, indeed than almost any of the three hundred brightest stars in our sky.

But it's not so much its own inherent qualities that make Aldebaran one of the few most famous and often observed of stars. Even more important are Aldebaran's many and wonderful relations to various stars, asterisms, clusters, constellations, planets, and the Moon. Those relations are a good place for us to start our study of the Bull's Eye.

The Most Connected Star

Surely no star is glorified more by its connections with other celestial objects than Aldebaran. To begin with, Aldebaran is the only 1st-magnitude star that lies right in front of a star cluster. And not just any star cluster: the closest and largest cluster that really looks like one, the Hyades. And the major stars of the Hyades form with Aldebaran a V shape that is a perfect depiction of the face of famous Taurus, the Bull—with Aldebaran as the Bull's bright eye. Not far west of Aldebaran and the Hyades is the more condensed cluster considered to be the loveliest of all to the naked eye—the Pleiades. The very name Aldebaran is Arabic for "the follower," a title that probably means the follower of the Pleiades—in their nightly journey across the sky.

Now, it's true that Regulus is closer to the ecliptic than Aldebaran and therefore more susceptible to close passes by the Moon and planets (not to mention the Sun—which passes closest north of Aldebaran around May 31/June 1 each year). But Aldebaran's huge accompanying clusters are like enlarged targets for producing glorious conjunctions that at least indirectly involve Aldebaran. Even if a planet gives a wide berth to Aldebaran during a particular year's passage, it goes between the Pleiades and Hyades, forming on a number of nights various striking patterns with the clusters and star, a chain of bright objects of which Aldebaran is a part.

Nor is it possible not to notice that Aldebaran is connected with Sirius through the link of Orion's Belt. The Belt points one way toward the brighter but lower Sirius, and the other toward the less bright but higher Aldebaran. Amazingly, the angular distances from the Belt to Aldebaran and the Belt to Sirius are virtually identical.

Another relational distinction of Aldebaran is that it's the star that leads the grand host of the winter constellations across the night sky and winter months. Orion is that army of brightness's brilliant center but Aldebaran and the forward-pointing arrowhead of the Hyades are its great vanguard (the gentle Pleiades float ahead as the banner bearer—or banner). Where do we begin the tracing of the Winter Hexagon and Winter Circle? Usually with foremost Aldebaran. It's certainly the star with which we begin the drawing of the other giant asterism of the brightest winter stars—the Heavenly G.

Aldebaran is connected to constellations not just by Orion's Belt, Sirius,

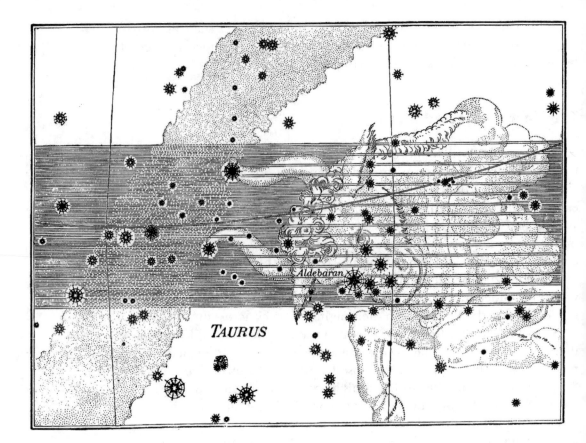

and the giant winter asterisms of bright stars. Extend the short "arms" of the
Aldebaran-and-Hyades-V straight out to the tips of Taurus's horns at Zeta
Tauri and Beta Tauri. The lower Hyades arm, the one ending with Aldebaran, is extended to Zeta Tauri (the star that the Crab Nebula is near) and
points onward to Gemini, connecting us visually with that constellation. The
upper Hyades arm is extended to 2nd-magnitude Beta Tauri (called Nath or
El Nath)—a star that quite literally connects to Auriga and Auriga's brilliant
star Capella. I say literally because the classic pentagon pattern that makes
Auriga recognizable requires using Beta Tauri. Taurus unofficially shares this
star of it with Auriga.

Occultations of Aldebaran

Along with Spica, Antares, and Regulus, Aldebaran is one of the four 1st-
magnitude stars that can be occulted by the Moon in our era of history. Alde-

baran is the brightest of those four stars, which makes it the second-brightest star that can ever be hidden by the Moon—the brightest is the Sun, during solar eclipses.

Aldebaran is therefore the brightest star that can be occulted by the Moon at night. And such events are lovely. The Moon hovers within the Hyades although, of course, if the Moon is at a large phase, it will be so bright you will need a telescope to see the Hyades stars properly and the cluster as a whole will spill well out of a telescopic field of view. The orangishness and point intensity of Aldebaran contrasts very beautifully with the yellow, duller Moon. And, pointlike as the star seems, you just may notice that Aldebaran isn't covered up quite instantaneously (as most stars seem to be). As we'll see later in this chapter, Aldebaran is one of the largest suns among the 1st-magnitude stars, subtending a larger angle than almost any other.

These lunar occultations of Aldebaran (and of the other 1st-magnitude stars that can be hidden by the Moon) occur in "seasons" usually a few years long within a longer cycle of lunar positioning. But sometimes your geographic location and cloudy weather may cheat you of seeing one of these events for many years. It took me almost forty years of having a good telescope before I finally got a chance to see an occultation of Antares. On the other hand, I was fortunate enough to see a few wonderful occultations of Aldebaran back in the late 1970s, but haven't been lucky with it since then.

My Aldebaran occultations may go a way back but another was observed almost exactly 1,500 years ago—and that sighting helped lead to a vital astronomical discovery.

The long-past occultation of Aldebaran was one that was observed in Athens in March of AD 509. More than a thousand years had to go by before Edmond Halley determined that for this event to have occurred, Aldebaran must have been several arc minutes farther north in ancient times. Halley compared the positions of Aldebaran, Sirius, and Arcturus in his own time with those that were noted in the distant past and found that these stars had all moved. In 1718, he announced his discovery of "proper motion"—movement of stars in the heavens caused by their true motions in space. In 2,000 years, the proper motion of Aldebaran adds up to about 7', which is almost one-quarter the apparent diameter of the Moon.

Aldebaran and the Clusters

Followers are generally deemed less important than leaders but, as important as the Pleiades are, their "follower" Aldebaran gives even greater emphasis and dignity to the pretty cluster. The converse is also true and about this, Martha E. Martin has written memorably in *The Friendly Stars*. She says that "Aldebaran shoots its ruddy face above the horizon just an hour after the hazy little dipper of the Pleiades has appeared, and the star is then to the east of, and, hence, almost directly under the Pleiades." Martin also remarks, "In his section of the sky Aldebaran reigns throughout all the lovely autumn evenings, with beautiful Capella in her own realm to the north of him and Fomalhaut far to the south . . ."

Aldebaran shines directly in front of the Hyades cluster and is a little less than 14° from the center of the Pleiades. These three entities—the star Aldebaran, cluster Hyades, and cluster Pleiades—are at significantly different distances from us in space. Aldebaran lies 65 light-years from us; the Hyades, about 150 light-years; and the Pleiades, maybe 390 light-years. So the Hyades cluster is about 2⅓ times farther than Aldebaran and the Pleiades cluster about 2⅔ farther than the Hyades.

The brightest Hyads shine at about apparent magnitude 3½ and have an absolute magnitude of 1 or 0 (a little less luminous than Aldebaran). As Burnham notes, there are four yellow giant stars of luminosity class III in the Hyades (Epsilon, Gamma, Delta, and Theta-1 Tauri) with the rest that we can see main sequence stars (luminosity class V) of spectral types A, F, G, K, and M. There are no hot young B stars in the Hyades, as there are in the Pleiades. The Hyades are maybe about 400 million years old, much older than the Pleiades.

The Lore of Aldebaran

Much of the lore of Aldebaran is shared by it with the Hyades or with Taurus as a whole. The heavenly Bull here seems to date back to long before Greek times. In fact, there is evidence that Taurus was originally the lead constellation of the zodiac and its formation may date back to around 4000 BC, when it marked the vernal equinox.

Taurus has most often been identified with the bull that Zeus became to carry away Europa. But Taurus is also a bull imagined to be challenging Orion and his hounds in the tableau of winter constellations that has come together from the work of many imaginations. Aldebaran's "rosy" color can be the bloodshot of the angry Bull's eye.

The name Hyades may be derived from the name of the brother of the Hyades maidens, Hyas, who was slain by a boar. Or could the derivation be the other way around? Some authorities have supposed that "Hyades" is from a Greek word meaning "to rain." Of course, the two ideas—Hyas and

rain—could be intertwined if we imagine that rainy seasons occur when the Hyades sisters are shedding tears for their dead brother. Certainly a rainy character was associated with both the cluster and Aldebaran, possibly because the Greeks experienced rainy seasons in late fall, when the cluster and star were rising at dusk, and in the late spring, when they were setting at dusk. A different but related connection might be alluded to in China's ancient work *She King*, for we read there that "the Moon wades through Hyads bright,/Foretelling heavier rain."

Aldebaran figures prominently in other lore that is independent of the Hyades or Taurus. One of the most stirring of such stories is told by the Hervey Islanders of Polynesia. They say that the Pleiades cluster was once a single star, the brightest of all, rivaling a half-moon in radiance. The Pleiad-star boasted of his beauty and was overheard by the god Tane. Tane became angry at the bragging and sought the help of Mere (Sirius) and Aumea (Aldebaran) to punish the star. The brilliant Pleiad-star heard the three coming and fled, first hiding under the Milky Way stream—until Mere

Aldebaran forms the V of Taurus's face with the Hyades. Note little dipper–shaped Pleiades near the center and Saturn bright at the lower right diverted it, revealing him. He fled again and this time would have escaped. But Mere grabbed Aumea—Aldebaran—and flung him so hard that Aumea struck the Pleiad-star, shattering the radiant light into six pieces. Tane was pleased, as was Mere (Sirius) who now became the new brightest of all stars. Aumea was glad to have his own light no longer drowned out by the nearby moon-bright star. But the six stars, now named Tauno, still gaze down—now forlornly—at their reflection in still waters.

By the way, it's interesting that the combined magnitude of all the Pleiades is really estimated as about 1.2 (another source says 1.5) and that of the Hyades as 0.5. So Aldebaran is about halfway in brightness between the total light of the two clusters. Yet surely the overall visual effect of the Pleiades—of multiform starry radiance almost filling more than a square degree of sky—is much greater than these numbers suggest.

Aldebaran in Modern Poetry and Science Fiction

Aldebaran makes some delightful appearances in poems of the past few centuries.

In the nineteenth century, American poet Lydia Sigourney wrote a poem called "The Stars" in which she advises us to "go forth at night,/And talk with Aldebaran, where he flames/In the cold forehead of the wintry sky."

If Aldebaran weren't the eye of Taurus, it could be located in the Bull's forehead. But speaking of the star as being in the forehead of the "wintry sky" is more interesting—it may, consciously or unconsciously, be a reference to that way Aldebaran leads the rest of the winter constellations across the sky (our aforementioned role of Aldebaran and the Hyades as the vanguard of the main army of splendor at whose front center Orion strides).

Another dramatic allusion to Aldebaran occurs in William Roscoe Thayer's poem "Halid": "I saw on a minaret's tip/Aldebaran like a ruby aflame, then leisurely slip/Into the black horizon's bowl."

Aldebaran has been a commonly mentioned location in the science-fiction and fantasy of the twentieth and twenty-first centuries. Robert Burnham Jr. wrote: "In the dark tales of his *Cthulhu Mythos*, H. P. Lovecraft . . . tells us that the Lake of Hali 'which is on a dark star near Aldebaran in the Hyades' is the present lair of the Old Ones, those mind-freezing, shambling horrors from beyond the barriers of space and time, who once held dominion over Earth, and seek ever to re-conquer it."

It's unclear whether the red star Borgil in J. R. R. Tolkien's classic novel *The Lord of the Rings* is Aldebaran or Betelgeuse.

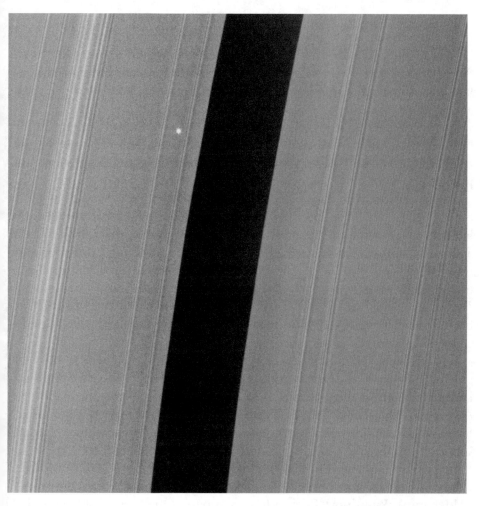

Aldebaran seen shining through a ring of Saturn as viewed by the Cassini space-craft.

The Departing Eye

When we learn of the true brightness and motion of Aldebaran in space, the story of the star in relation to us is as thought provoking as any of the myths and legends about it.

Ever since I was quite young, I was impressed by a certain seldom-noted fact about Aldebaran: it is receding from us much more rapidly than is any other 1st-magnitude star. In fact, of the more than three hundred stars that are brighter than magnitude 3.5, I count only three that are moving away from us at a greater rate than Aldebaran. Aldebaran is receding from us at a speed of 54 km/sec.

Aldebaran was obviously once a lot closer to us than it is now. According to University of Texas astronomer Jocelyn Tomkin, Aldebaran was the brightest star in Earth's night skies in the period from about 420,000 to 210,000 years ago. This was perhaps the time when humans of our particular species first came into being. Aldebaran was closest to us and brightest about 320,000 years ago. At that time, the orange giant was only 21.5 light-years from Earth—a little less than a third as far as it is now. The maximum brightness of Aldebaran back then should have been –1.54. That is slightly brighter than Sirius is today and brighter than any other star in about the past million years.

In the period from about 950,000 to 1,370,000 million years in the past, two A-spectral-type stars, Zeta Leporis and Zeta Sagittarii, passed respectively 5.3 and 8 light-years from Earth, achieving maximum magnitudes of –2.05 and –2.74. Before that, Canopus ruled for over 2 million years with a brightness that peaked at –1.86. And in the 1.3 million years before Canopus, back to 5 million years ago, two B-type stars, Beta Canis Majoris (Mirzam, the "announcer" of Sirius) and Epsilon Canis Majoris (Adhara) burned as brightly as –3.65 and –3.99, respectively, when they passed 37 and 34 light-years from Earth.

In the *future* 5 million years, only two stars—Sirius and Delta Scuti—will get brighter than Aldebaran did. But it's the star just after close-passing Delta Scuti's short reign that is most interesting to consider in relation to Aldebaran. That star is Gamma Draconis, also known as Eltanin. Eltanin is a K5III star like Aldebaran. But then so is Earth's brightest star after Eltanin, Upsilon Librae. The thing is, Eltanin is much more similar to Aldebaran than Upsilon Librae is. The absolute magnitude of Eltanin is –1.1, compared to –0.8 for Aldebaran. Their B-V color indices are +1.52 (Eltanin) and +1.54 (Aldebaran). Right now, Eltanin has an apparent magnitude of 2.24 and is 148 light-years away. But from 1,330,000 to 2,030,000 years in the future, it will be Earth's brightest nighttime star, shining at a peak magnitude of –1.39 when it passes 27.7 light-years from us (remember, Aldebaran's closest pass was 21.5 light-years, its greatest brightness –1.54).

The final remarkable similarity of Aldebaran and Eltanin, though just a coincidence of human pattern-making, is that both may be considered as marking the eye of the constellation to which they belong. Aldebaran is the Eye of Taurus, the Bull. Eltanin may be considered the Eye (or one eye) of Draco, the Dragon. So our solar system is being watched by a dimming eye—Aldebaran—as that eye recedes and by a brightening eye of almost identical color, spectral type, and luminosity—Eltanin—as this other eye approaches.

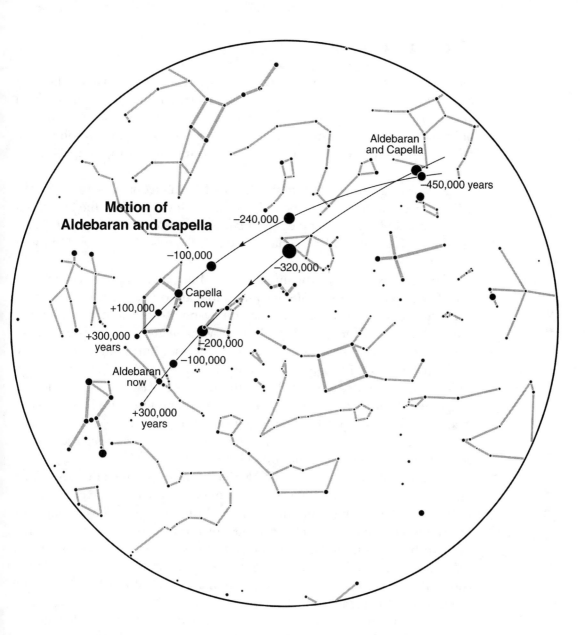

The paths of Aldebaran and Capella through the heavens over a long period of time, including their close pairing a little less than 450,000 years ago (note that the background stars would move, too—but most not as far as our close neighbors Aldebaran and Capella).

Aldebaran as a Sun

We've already discussed some of the properties of Aldebaran as a sun—that is, as a star considered close up as a roughly spherical body bringing heat and light possibly to a family of planets or other "solar system" bodies of its own. But there's much more to say. And before we're done, we'll consider not only Aldebaran as a sun but our Sun—in the future—as an Aldebaran!

Aldebaran is a K5 star of luminosity class III, which qualifies it as a "giant." The surface temperature is about 4,000 K, much cooler than the Sun's. But Aldebaran is about forty times the diameter of the Sun (at 65 light-years away, that gives an angular diameter of 0.02"), so it is 350 times as luminous as the Sun.

There seems to be a surprising amount of uncertainty about key properties of Aldebaran—at any rate in sources that are otherwise usually helpful. For instance, estimates of its mass seem to range from about 1 to 2½ times the Sun's mass. Furthermore, Aldebaran is variable in brightness but sources don't seem to give an average range or any period. Burnham states that the maximum recorded range of brightness is "about 0.78 to 0.93." Kaler says Aldebaran fluctuates erratically by about two-tenths of a magnitude in brightness. In addition, some sources seem to suggest that Aldebaran has no true companion star, while others mention a 13th-magnitude red dwarf that shares Aldebaran's proper motion. This star is about 0.5' from Aldebaran, with a true separation of something like 600 AU.

There is also speculation about Aldebaran having either a massive planet or brown dwarf circling it. One study seemed to reveal an extremely slight wobble in Aldebaran's motion, which could be interpreted as a planet or brown dwarf with eleven times the mass of Jupiter circling Aldebaran in a two-year period. But no later study has confirmed this. And, interestingly, it is believed that Aldebaran's rotation period is almost two years. That's roughly 25 times longer than the rotational period of our Sun and 2,500 times that of the Pleiades star Pleione.

Most interesting of all is the possibility that when we look at Aldebaran we are seeing just about what our Sun will look like roughly 5 billion years in the future. How close the resemblance will be depends on whether the mass of Aldebaran is very similar to that of the Sun or is as much as 2½ times greater. Our Sun will definitely someday become even *more* luminous and large than Aldebaran now is. But Aldebaran is still increasing in luminosity and size. It already has run out of hydrogen and is being powered by the conversion of helium into carbon.

The View from Aldebaran

An observer in the Aldebaran system would not see the Hyades tremendously brighter than we do and they would be scattered over a fairly large section of sky. Whether or not the red dwarf about 660 AU from Aldebaran is truly traveling through space with it, it should appear as a brilliant object in the sky of any planet near Aldebaran. Oddly, there seem to be few stars of respectably great luminosity near Aldebaran. You have to travel about 14 and 15½ light-years from the star before you encounter two stars that go by even Flamsteed designations (quite dim to the naked eye for us): 39 Tauri and 104 Tauri, both G-type stars. How far do we have to go from Aldebaran before we reach stars that are 1st- or 2nd-magnitude in Earth's sky? Capella is 36 light-years from Aldebaran and Hamal (2nd-magnitude ornage Beta Arietis) is 39½ light-years. Castor and Pollux are, respectively, about 45 and 48 light-years from Aldebaran.

The Ultimate Double North Star

Capella may be fairly far from Aldebaran in both space and the sky in our time. But the figure on page 205 reveals what might be the most amazing relationship ever to have occurred between two of our brightest stars—one that would have produced possibly the greatest visual show. The diagram plots the paths of Aldebaran and Capella 300,000 years into the future but, more interestingly, 450,000 years into the past. The first thing you notice is that not quite all the way back, there was a crossing of the paths of the two stars and a remarkably close pairing of Aldebaran and Capella when both were bright. Note that Aldebaran especially was bright then—much brighter than the symbol for Vega at its current position not far from the ancient Aldebaran-Capella meeting point.

But notice also that Aldebaran and Capella were almost as close together for tens of thousands of years. They were together long enough for precession of the north celestial pole to pass near them. When the pole did, Earth had not just a brilliant North Star but a double brilliant one.

22 SPICA

Spica has long been one of my favorite stars. My extra affection for it doesn't stem quite all the way back to when I first identified it (when I was a very young child). Instead, the added dimension to my appreciation of Spica surely began on a wondrous night in 1969. That was the night I observed the star extremely near a totally eclipsed Moon—the first lunar eclipse I'd ever seen in a telescope.

I'll have more to say about that rare event as well as beautiful occultations of Spica by the Moon that I have observed. But there's much more of interest about Spica than just its susceptibility to meetings with solar system bodies. Spica is the one bright star in the longest constellation of the zodiac, Virgo. It is a star that is usually rather lonely, ruling over a sizable region of the heavens that is fairly dimly and sparsely starred. But Spica has connections to other bright stars and star patterns—connections that are strong by virtue of their geometry and the fame and brightness of the other sights. And science of recent decades has revealed Spica to be one of the most physically remarkable of the brilliant stars.

Spica's Connections with Bright Stars and Constellations

Although Spica's constellation is dim, some compelling geometric links connect it to bright stars and other constellations.

The most famous is one we've already discussed in this book: the arc from the Big Dipper's handle to Arcturus that is continued by the "spike" (straight line) from Arcturus to Spica. The modern pronunciation of Spica is *SPY-kuh*, even though the original Latin pronunciation of the name would supposedly be *SPEE-kuh*. A rare alternative to "drive a spike to Spica," which would favor the ancient pronunciation, is to "speed on to Spica" from Arcturus.

The straight-line distance from the end of the Big Dipper's handle to Arcturus is about 31°; that from Arcturus to Spica is just less than 34°. I sometimes say that after "taking the arc to Arcturus and driving the spike to Spica" an observer can "continue the curve to Corvus." Corvus, the Crow, is

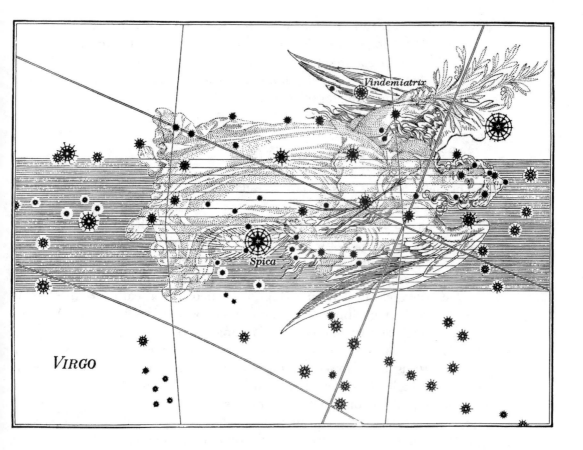

actually rather close to Spica. The middle of Corvus's conspicuous rhomboid of stars is only about 17° from Spica. Indeed, one way for a beginner to confirm that Spica is Spica (especially if Arcturus and the Big Dipper are hidden by clouds or higher tree branches) is to look for Corvus to the right of Spica (right as seen from the Northern Hemisphere). Corvus is actually somewhat south of due west of Spica, so the two are at about equal height when they are in the southeast, Corvus leading.

More often, the identification function between Corvus and Spica works in the opposite direction: you may be observing with rather bright moon or haze or light pollution and have to use bright Spica as a guide to finding Corvus.

But these are just the beginning of the Spica connections. If you draw a line about 49° almost due south, it will take you to the center of a pattern more compact and very much more prominent than Corvus: the Southern Cross. Of course, if you live north of about 30° N latitude, the Southern

Cross doesn't get above your south horizon. But it's still exciting to use Spica as a guide to judging exactly where below your horizon the legendary cross shines. If you live a little farther south, where Crux is going to be so low in the sky that it will be greatly dimmed, Spica can be a wonderfully practical guide to helping you glimpse its stars.

Spica and the Diamond of Virgo

Now Spica is a little closer to Crux, Beta Centauri, and Alpha Centauri, than it is to Regulus. Regulus is 54° from Spica—farther than Antares (46°) is from Spica. The bright form of Leo also interferes with any natural inclination to try forming a vast "Spring Triangle" of Spica with Regulus and Arcturus. Instead, Spica and Arcturus are linked with the stars Denebola (Beta Leonis, the tail or rump of the Lion) and Cor Caroli (the Alpha star of Canes Venatici, the Hunting Dogs, crouched underneath—south of—the handle of the Big Dipper). This pattern is known as the Diamond of Virgo—even though only one of its four stars, Spica, is part of Virgo.

The Diamond of Virgo really is an interesting and, as it turns out, very useful asterism. The lower two-thirds of the Diamond is a nearly perfect equilateral triangle: the distance from Spica to Arcturus is just less than 34°, from Arcturus to Denebola is 35°, and from Denebola to Spica is also 35°. The top part of the diamond is considerably smaller: the side of it from Arcturus to Cor Caroli is about 26°, from Cor Caroli to Denebola is 28°.

I said that the Diamond of Virgo is useful—and I meant for locating objects in this part of the heavens. The heart of the amazing Virgo Galaxy Cluster is wedged within the western angle of the Diamond (the angle whose point is Denebola). About half way along the side of the Diamond that runs from Denebola to Cor Caroli is the giant loose naked-eye star cluster in Coma Berenices called the Coma Cluster. About two-thirds of the way along a line that bisects the Diamond—the line from Denebola to Arcturus—you can find 4.3-magnitude Alpha Comae Berenices and about 1° northeast of it a 7.7-magnitude globular star cluster, M53. And these are just a few of the interesting deep-sky sights we can use the Diamond to locate.

The Virgin's Ear of Wheat

Spica's constellation is Virgo, the Virgin. Virgo is the mysterious maiden who is pictured reclining languorously with the symbol of her patron art of agriculture in her left hand at her hip. That symbol is an ear or sheaf of grain, usually said to be wheat—and it is marked by Spica. The name Spica literally means "ear of wheat."

Who is this lady who holds Spica? The idea of this constellation's being an agriculture goddess holding grain seems to derive from far back in early Mesopotamian cultures. In classical times, Virgo was often identified with the Greek goddess Demeter, whose Roman version is Ceres—goddess of agriculture and the growing world. We get the English word *cereal* from Ceres. And the name Ceres was also given to the first-discovered and largest of the asteroids. (Ceres, about 600 miles across, was in 2006 categorized as a "dwarf planet" by the IAU—though it remains to be seen whether this controversial designation will endure for Ceres any more than it will for Pluto.)

Virgo is visible in the evening sky in the spring—appropriate for a goddess of green, growing things. But the Sun is in this constellation around harvest time (in the early twenty-first century, from September 17 to October 31—but about a month earlier in classical antiquity, due to the precession of the equinoxes). This might have been a more important link between Virgo and Demeter/Ceres in the minds of ancient Greeks and Romans (though I confess I have little idea of exactly when their most important harvests were in various parts of Greece and Italy over the centuries these cultures flourished).

I said the figure of Virgo is mysterious. One reason for this is her wings. Depictions of her with wings go far back—but this would not be consistent with her being Demeter/Ceres. Some have suggested that she is Nike, goddess of victory, also memorialized in one of the most famous statues of antiquity. Others have argued that she is the Greek goddess Astraea, "last of the celestials to leave the earth, with her modest sister Pudicitia, when the Brazen Age began." Could Virgo also be the blindfolded Roman goddess of Justice because Libra, the Scales, is the next constellation of the zodiac after her?

An image that has occurred to me is that Virgo's head is surrounded by a cloud of innumerable glowing dreams—the galaxies of the great Virgo Cluster.

The Names and Lore of Spica

In ancient Egypt, Spica was supposedly very important, being known as the Lute-Bearer and Repa, meaning "the Lord." The English astronomer Sir Joseph Norman Lockyer believed that several very important temples of very ancient Egypt and ancient Greece were dedicated to Spica, which may have been associated with the legendary Egyptian personage Menes.

In medieval times, one of the names that the Arabs applied to Spica was Sunbulah. The nineteenth-century deep-sky writer Admiral Smyth states that, before his time, there was an attempt to rename the star Newton, after the great scientist. But perhaps the most enduring alternate name of Spica is Azimech. That name comes from the medieval Arabic title Al Simak al A'zal, which means "the Unarmed Simak." Arcturus was the other Simak, the Armed one, Al Simak al Ramih. R. H. Allen speculated that *simak* was from a root that meant "to raise on high." But our contemporary star-name expert Paul Kunitzsch states he feels that the meaning of *simak* is uncertain. We do know that the part of the two stars' titles about being armed or unarmed is probably a reference to the fact Arcturus has some modestly bright naked-eye stars near it and Spica doesn't. Eta Boötis (Muphrid) and a few other stars might be the weapon of Arcturus, a lance that made it the "lance-bearer."

Spica and Arcturus are paired in another memorable imagining, this one in ancient China. According to Gertrude and James Jobes, the two bright stars were the Horns of the Dragon and spring was calculated from the time that the Full Moon appeared between these horns. "With warmth of feeling, a great display of joy," write the Jobeses, "the people, who used the skies as a calendar, watched the winter draw to a close when the Moon 'rode the Dragon's Horns.'" Furthermore, they say, "This association with spring may have been the reason Show Sing, god of long life, may have chosen Spica for his home. This venerable and wise old man rode about on a stag accompanied by a bat, symbol of happiness and longevity. Show Sing, smiling and kindly, always carried peaches, the fruit of immortality."

Occultations and Conjunctions of Spica

In some years, the Full Moon of early spring can't quite fall between the horns of the dragon, Arcturus and Spica, because it passes a little south of Spica. The Moon more often goes north of the star. Sometimes, however, there is a lunar occultation of Spica. Using a telescope, I've seen Spica emerge from behind the Moon as the latter climbed the southeast sky a while before sunset. The star was fragile but strong in the day sky, trembling just beyond the bright edge of the gibbous Moon. Even more beautiful through a telescope was the sight I had years earlier of Spica coming out from behind the dark side of the Moon in an exquisitely clear sky right around sunrise. As the minutes after sunrise passed, I continued to be able to see the star in my 8 × 50 finderscope. The bright spark of Spica twinkled intensely in my one eye as my other eye saw myself and the landscape flooded in strong sunlight. I was being treated to potent views of two of my favorite stars—our Sun and Spica—at one and the same time.

Yet a far greater experience involving Spica came for me way back on the night of April 12/13, 1968. Although I had seen a few lunar eclipses with my naked eye before, this was my first chance to view one in detail with a telescope. I recall the night being wonderfully clear and I know I observed from a little hill on the north edge of our property. My interest in seeing this eclipse had been further inflamed by the local newspaper's little syndicated box on astronomy for that day. For my geographic region, the box was called "Atlantic City Skies" (it was named after the city of the newspaper it appeared in). It gave the time of local sunrise, sunset, moonrise and moonset, but also, for each day, a statement about some astronomical event or sight that was occurring. For this day it noted that the totally eclipsed Moon would be appearing very close to the star Spica, an event not to be repeated for an extremely long time—I think the claim was for a few hundred years.

All I know is that not just the telescopic view of the eclipsed Moon, ruddy but rimmed with yellow and blue, was beautiful but the also the naked eye view of the reddened Moon hanging over my southern tree line. In fact, the latter sight was extraordinarily, unforgettably beautiful because as more and more of the Full Moon was shadowed, stars came out ever more thickly in the sky and, less than a degree lower right from the Moon, Spica kindled ever stronger next to the softly glowing giant ember. This was one of the great chances in my life for the beauty and power of a star to be so enhanced by its setting that it was revealed to me in all its primal wonder. I remember sharing some of this experience of splendor with my mother, who perhaps appreciated it as much as I did.

Spica is involved in numerous impressive conjunctions with Moon and planets over the years. Spica and Antares were the only 1st-magntude stars that were part of the astounding gatherings of all the planets visible at once that occurred in February 1982 and January 1984—and will not occur again for centuries (see the chapter "All the Worlds in My Window" in my book *The Starry Room*).

If you are interested in rare celestial events, one of the rarest is the occultation of a 1st-magnitude star by a planet. Spica and Regulus are apparently the only 1st-magnitude stars that were occulted by planets in the past few thousand years or will be in the next few thousand. According to one calculator, the last occultation of Spica by a planet was one by Venus on November 10, 1783, and the next, again by Venus, will not occur until September 2, 2197.

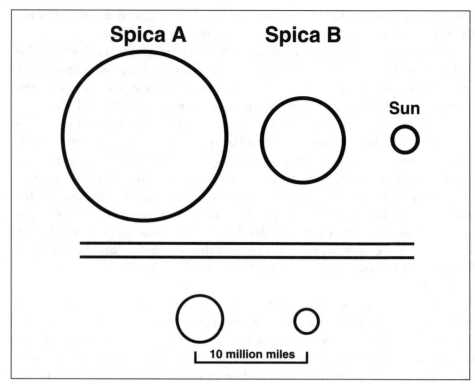

Relative sizes of Spica A, Spica B, and the Sun (top diagram); relative sizes and separation of Spica A and Spica B (bottom diagram).

Spica as a Sun—or Suns

Five of the nineteen brightest stars are objects with spectral types between B0 and B3—bluer and hotter than Rigel or Regulus. These stars are Achernar, Beta Centauri, Alpha Crucis, Spica, and Beta Crucis. Thus four of the "South Hemisphere Six" are stars of this kind and only one 1st-magnitude star of this kind—Spica—is visible to most of the world's people, those who live around 40° N.

Actually, several of these stars consist of two or more hot, bright component suns. Spica is an example. The brighter of its two major components has a surface temperature of 22,400 K, a luminosity 13,400 times that of our Sun, and a diameter about 7.8 times that of our Sun. The second component of Spica has a temperature of 18,500 K, 1,700 solar luminosities, and 4.0 solar diameters. The larger, more luminous star has a mass about 11 times solar and may therefore someday go supernova. The lesser star has a little less than 7 times the mass of the Sun.

What's really fascinating is that the two big, bright, massive components of Spica are so close together. Their center-to-center separation averages just

0.12 AU—that's about 11 million miles, with the larger star having an average diameter of nearly 3½ million miles. Notice the words "averages" and "average" in the previous sentence. The distance between the two varies a bit because the two go around each other in a slightly elliptical orbit. The larger star—and the smaller one, too—have an average diameter because they are close enough to each other to tidally distort each other into slightly ellipsoidal shapes. Astronomers used to think that Spica's variations in apparent brightness were caused by very slight (grazing) eclipses of one of the component suns' disks by the other as seen from Earth. Now they believe that the changes in apparent brightness are caused by us being presented with varyingly wide diameters of these slightly ellipsoidal suns. Spica's apparent magnitude varies between +0.92 and +1.04 with a period of just over four days—the amount of time it takes for the pair to orbit each other once!

An additional, very small fluctuation in Spica's brightness is caused by the fact that the primary (brighter, larger star) is a pulsating variable of the Beta Cephei (or Beta Canis Majoris) type. This variation is only 0.015 magnitude in a period of just 0.17 days.

Spica is hot enough to produce huge amounts of ultraviolet radiation. But it is also a strong X-ray source, probably in large part because of collisions of the solar winds of the two stars.

Could there be more stars in the Spica system? Apparently there is evidence from lunar occultations that there are three other, fainter components.

ANTARES 23

On a clear evening in July, the grand band of the softly glowing Milky Way widens and brightens as our eyes wonderingly follow it down the southeast sky. But just ahead of this lowest, brightest, broadest part of the band blazes a long twisted line of fire—a curve of stars radiant enough for its shape to seem to be almost branded upon the sky. This pattern is Scorpius, the Scorpion, brightest constellation of summer and brightest of the zodiac. It wouldn't be Scorpius without its most outstanding star, the 1st-magnitude star that burns midway along its fallen-forward S, flickering deep orange-gold: Antares.

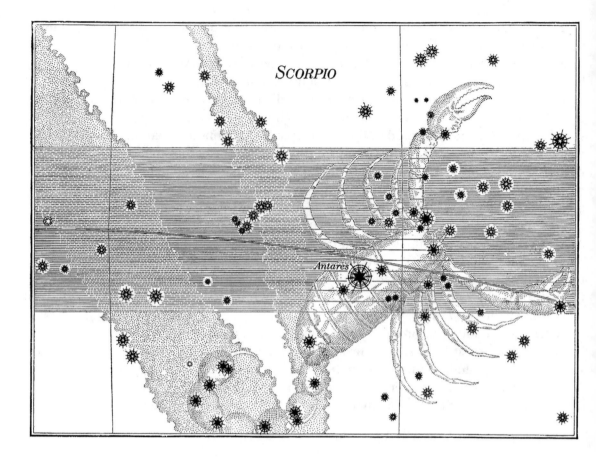

SCORPIO

Antares

Scorpius and
Antares.

Antares: famed for being the fiery heart of the Scorpion; named for its color and rivalry with Mars; renowned as the great red giant star of summer, possibly the superior of even winter's Betelgeuse in some respects. Antares—the burning coal deep in the southern heart of northern summer.

The Heart of the Scorpion

Antares resembles Altair in having a moderately bright star fairly close to either side of it. But in the case of Antares, the flanking stars are almost the same in brightness. Tau Scorpii, southeast of Antares, shines at magnitude 2.8; Sigma Scorpii, northwest of Antares, shines at magnitude 2.9. Together, these two stars have been called the Praecordia—"the outworks of the heart."

An even more interesting neighbor of Antares in the sky is the globular cluster M4. M4 is positioned only 1½° to the west of Antares and closer to the southeast of Sigma Scorpii. In space, the cluster is very much farther from us than Antares but is actually one of two rivals for the title of closest globular cluster to Earth (the other is NGC 6397 in the far-south sky in Ara). M4 may be a little less than 7,000 light-years from us. No wonder then that, through telescopes, it looks much larger than your typical globular cluster and that its individual stars are much easier to resolve. Viewed through very large amateur telescopes, M4 becomes one of the most stunning of all deep-sky objects, displaying a spectacular "bar" of stars across its center.

The Rival of Mars

It is appropriate for a star marking a heart to be somewhat ruddy. But the color of Antares is so fascinating that the star's very name is derived from its hue. The name Antares means "the rival of Mars." But this is not a reference to Antares's giving Mars competition in brightness per se. It alludes to the similar color of the two.

Let's consider the etymology of the name. Ares was the bloody Greek god of war, whose Roman equivalent was Mars. The Greek name for the star is literally "anti-Ares"—against Ares. The battle is one of redness. The B-V color index of Antares and the average B-V color index of Mars are very similar. I say "average" for Mars because widespread dust storm activity on the planet can reduce the ruddiness of the naked-eye point of light that is the "Red Planet." Many observers have also noted that when Mars gets unusually bright, its ruddiness seems lessened. My idea about this is that when Mars gets brighter than about –1, its brightness begins to wash out the color. To the color receptors in *my* eyes, at least, Mars may be ruddiest when it's between about magnitude +0.5 and –0.5, or a bit brighter. Antares is a little fainter than the dim end of that range and so, in my opinion, never appears as red as Mars at Mars's reddest. On the other hand, Mars spends most of its time far enough from Earth to shine dimmer than magnitude 1. Mars is usually dim enough to appear slightly less ruddy than Antares.

When we talk about Antares or Mars appearing "ruddy" or "red," we should always be careful to note that the real hue of these lights is not so extreme. A better term for this color may be orange-gold. I've referred to Mars as being tiger-colored, campfire-colored, and also pumpkin-colored. The last of these might be best, especially since pumpkins can display varying degrees of gold or orange, depending on how ripe they are.

The color of Antares has been noted with delight by many famous observers. A classic description of the hue comes from Martha Evans Martin in *The Friendly Stars*: "I am not sure that the color of Antares is any reason why it should be associated with the blooming of red flowers but I find in my journal that I have unconsciously made the association. One entry, on June 30, says, 'Antares is shining splendidly tonight and rivals in color the wild red lilies that were blooming today in the coppice down beyond the spring.'" Evans also made a famous statement that the star's "redness is still further intensified by the white sweep of the Milky Way, which is now very bright, and which lies within a few degrees of Antares."

The Green Companion of Antares

There is one final influence on an observer's perception of Antares's color, though it only holds during a particular telescopic observation. I'm referring to seeing Antares seem ruddier in relation to the apparent *green* of its companion.

According to Robert Burnham Jr., the companion to Antares was probably first seen by one Prof. Burg at Vienna on April 13, 1819—during an occultation of Antares by the Moon. Burg witnessed a star he estimated as magnitude 6.7 emerge from behind the dark part of the Moon and then suddenly turn, about 5 seconds later, into a 1st-magnitude light. He correctly interpreted what he saw as proof that Antares was a double star whose much-dimmer component was normally overwhelmed by the brilliant primary.

You don't have to wait for an occultation of Antares to glimpse the companion. Garrett Serviss, in his *Pleasures of the Telescope*, wrote:

> Antares carries concealed in its rays a green jewel which to the eye of the enthusiast in telescopic recreation appears more beautiful each time he penetrates to its hiding-place. . . . When the air is steady and the companion can be well viewed, there is no finer sight among the double stars. The contrast in colors is beautifully distinct—fire-red and bright green. The little green star has been seen emerging from behind the moon ahead of its ruddy companion.

A telescope as small as 4¼ inches can reveal Antares B and Burnham comments that he found this star "fairly easy" in the Lowell Observatory 7-inch refractor "*when the air is calm*" (emphasis added).

Good "seeing" is the key. In the United States, and surely many other countries, summer often brings hazy but steady skies, so nights when one can split Antares and see the companion should not be too rare. The companion is not all that dim. Despite fainter estimates by early observers of Antares B, modern measurements show the star to be magnitude 5.4. Whereas the companions of Sirius and Vega are about 10,000 times fainter

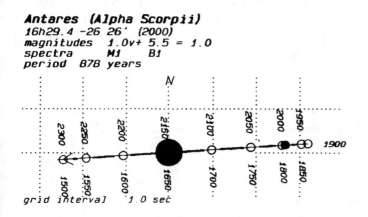

Antares (Alpha Scorpii)
16h29.4 -26 26' (2000)
magnitudes   1.0v+ 5.5 = 1.0
spectra   M1   B1
period   878 years

N

2300 2250 2200 2150 2100 2050 2000 1950 1900

1500 1550 1600 1650 1700 1750 1800 1850

grid interval   1.0 sec

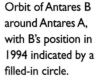

Orbit of Antares B around Antares A, with B's position in 1994 indicated by a filled-in circle.

than their primaries, Antares B is only about 60 times fainter than its primary. On the other hand, it is very much closer to its primary than Vega's companion and somewhat closer than Sirius's. Antares B was 2.65" from the primary star at the start of 2008 and the separation is currently shrinking by only about 0.01" a year (we'll discuss more about the true separation of the primary and secondary in space, and the secondary's orbit and physical nature, later in this chapter).

The big immediate question about the companion of Antares is whether it *really* appears green.

There are certainly a lot of us who see it as distinctly—even flagrantly—green when in the field of view with Antares itself. In addition to Serviss, William Tyler Olcott refers to the companion of Antares as being "vivid green." Mary Proctor called it "the wily companion of verdant hue." E. J. Hartung writes that the color of Antares B is "pale green which I have seen well with 30 cm in bright sunshine against the blue sky." Is the star truly green, though? Couldn't the green be explained as being entirely a contrast effect, the orange of the primary inducing us to perceive a complementary green in the companion?

The true test of the companion's color should be what it looks like when it emerges from behind the dark side of the Moon before the emergence of Antares. Fortunately, the companion lies almost exactly due west of the primary and so comes out into sole visibility for typically about five or six seconds before Antares bursts back into view. Do observers then see it as green?

The answer seems to be a qualified yes. At least some observers see it as distinctly greenish even when Antares itself is still hidden by the Moon. Back in 1856, one of the greatest double-star observers of all time, W. R. Dawes, watched the companion emerge from behind the Moon and thought it looked greenish. Almost 150 years later, on March 3, 2005, another skilled observer, Steve Albers, saw the emergence of the companion at 60× in a 6-inch Newtonian:

> The reappearance was a little like seeing Ganymede coming back prior to Jupiter's reappearance [from a lunar occultation] in the fall that I had seen in binocs. This time though Antares B looked much brighter than I expected, almost making me think it was Antares itself. And I could make myself believe that it had a greenish tint. Then after the all too brief 5 seconds, the big fire returned, and the companion regained its usual status of being invisible. The briefness of this could be compared to seeing the diamond ring at a solar eclipse perhaps.

When quizzed further, Albers said he felt the green was definite, not subjective, and also noted that the hue was more bluish-green than yellowish-green.

It is rare to get a really good opportunity to see the companion of Antares alone at a lunar occultation. But consider how rare it is for Antares to be occulted by a planet. The next such event is an occultation of Antares by Venus, which will occur on November 17, 2400. And the previous occultation of Antares by Venus? That happened almost three thousand years earlier.

The Rival of Betelgeuse

Let's turn at last to the matter of what Antares is like as a sun. We've already discussed the other 1st-magnitude star that is a red giant, Betelgeuse. Perhaps the best way to begin our profile of the physical nature of Antares is therefore by comparisons to Betelgeuse.

Until recent years, such comparisons would not have been flattering to Antares. Betelgeuse was regarded as the more luminous, larger, and cooler star (which is more impressive at this cool end of the range of spectral types). Betelgeuse is almost always a star of brighter apparent magnitude than Antares (both stars are variable in brightness—more on this in a moment). But which star has the greater true brightness—and true diameter—depends on how far they are from Earth. That remains poorly known because these vast, diffuse-edged, pulsating stars are difficult to get accurate enough positions for to allow good parallax measurements. But in recent years, most authorities seem to estimate Antares as being considerably farther from us than Betelgeuse—and therefore having greater true brightness and size. For instance, one source has Betelgeuse as 520 light-years away and therefore an average absolute magnitude of about –5.0, but places Antares 600 light-years away with an average absolute magnitude of –5.8. Another source, older and perhaps less reliable, lists the distance to Betelgeuse as only 350 light-years (a mere 40 light-years farther than Canopus) and to Antares as 450 light-years. If these latter, closer distances to us are true than Betelgeuse would have an absolute magnitude of –4.9 and Antares, –4.8.

James Kaler emphasizes the uncertainty in the distances of the two stars but opts to give a distance of about 425 light-years for Betelgeuse and 600 for Antares. The diameter of Betelgeuse in visible light, he gives as about 2.8 AU, and he lists two different methods for determining Antares's diameter—one that gives a figure of about 3 AU; and the other, about 3.8 AU. And yet he seems to list the same luminosity—60,000 times solar if we include the large percentage of infra-red light—for both stars. Could this equality be due to the fact that Betelgeuse is a subtype M2 star and Antares an M1.5, so that Betelgeuse, though smaller and less luminous in visible light, is cooler and produces more infrared radiation than Antares? Actually, among M-type stars, scientists think that surface temperature does not have a strong correlation with subtype—an M2 star is not necessarily cooler than an M1.5. Kaler, in fact, lists the surface temperatures of Betelgeuse and Antares as both being approximately 3,600 K. So on this point, too, the jury is out.

Betelgeuse and Antares are both luminosity class Iab supergiants. Kaler suggests that the initial mass of Antares was about fifteen to eighteen times that of the Sun, whereas that of Betelgeuse was about twelve to seventeen solar masses. Both stars are almost certainly destined to end up as supernovae. Both emit radio energy; both appear to have vast sunspots and bright regions upon them. And both vary considerably in both size and brightness.

One category in which Betelgeuse does seem to exceed Antares is in its range of variability. Betelgeuse fluctuates in brightness by perhaps about 0.4 magnitude commonly, and has a rare more extreme range of perhaps more than 1.5 magnitudes. The variations of Antares are rarely more than from about magnitude 0.86 to 1.06, and the star apparently never gets brighter than this most typical range. But an extreme minimum of 1.8 has been reported for Antares. At this point it would be only the second (or, if Delta Scorpii were in its bright state, third) brightest star in Scorpius. Could Antares get even dimmer on rare occasions? Eratosthenes said that Beta

Librae was the brightest star in the combined Scorpion (Scorpius) and its claws (Libra). Ptolemy held that Beta Librae and Antares were of the same brightness. Today, Beta Librae shines at only magnitude 2.6, dimmer than several stars of Scorpius. So if we correctly understand those ancient authorities, and if they were right—two big ifs—then Beta Librae must have been considerably brighter than it is now, not just Antares dimmer.

The main brightness variation of Antares has a semiregular periodicity of approximately 1,733 days, or 4.75 years.

Antares's Surroundings and the Companion's Nature

Both Antares and Betelgeuse are surrounded by nebula—very difficult to perceive visually, though in the case of Antares not that difficult for a skilled astrophotographer to record. This reddish nebula is about 1° wide or roughly 5 light-years across. Despite its color, the Antares nebula apparently shines by reflected, not emitted light—the light from Antares is reflected off fine, solid particles rather than gas. An area of dark nebula starting just north of Antares extends for about 4° north where it meets long lanes of dark cloud that extend more than 10° to the east and northeast. One of these spooky lanes is the Pipe Nebula, detectable with the unaided eye and slight optical aid, which forms a back leg of the Great Galactic Dark Horse, a figure typically only traceable on imaging of this splendid Sagittarius–Scorpius region of Milky Way. There is luminous nebulosity in this area of the heavens, too, particularly about 3° north–northwest of Antares, where it surrounds the double star Rho Ophiuchi. As Burnham notes, the dark area just west of the Rho Ophiuchi Nebula is probably the "hole in the heavens" that Sir William Herschel mentioned in 1785.

What about much closer to Antares, where its companion orbits? The companion lies roughly 550 AU from Antares and is thought to be circling it in a vast, slow orbit. Many different estimates of the orbital period have been given, but a recent one I find is 1,218 years. On the other hand, James Kaler gives the period as about 2,500 years. However long the orbit takes, we know that the companion has hollowed out an ionized region within the solar wind from Antares. It is a hot star of spectral class B2.5—therefore it ought to look blue-white. Might its light be rendered more greenish by passing through the (dim but particle-filled) Antares Nebula? The companion is at least hundreds of times more luminous than our Sun. Its mass is about seven to eight solar masses. After Antares goes supernova—sometime between tonight and a few million years from now—the companion will live on and probably never go supernova itself. Its fate will be to become a red giant itself someday and probably end its life as a massive white dwarf.

Antares in the Scorpius-Centaurus Association

Antares is believed to be the most massive, luminous, and evolved star in the great Scorpius-Centaurus Association. This association consists entirely of young, hot, luminous stars (that are almost exclusively of type B)—except for Antares, which is massive enough to have already progressed to the red giant phase. To judge from the motion and expansion of the Scorpius-Centaurus Association, its stars, all born around roughly the same time, must be less than 20 million years old. The association contains well over a hundred stars, spread out over about 90° of our sky. The great span of this formation is due to the fact that it is the closest of all stellar associations. It is centered roughly 550 light-years from Earth, near Alpha Lupi and Zeta Centauri. This places it more than twice as close as the even more majestic Orion Association. Among the most prominent stars in the Scorpius-Centaurus Association are Antares, Beta Crucis, Sigma Scorpii, Epsilon Centauri, Alpha Lupi, Delta Scorpii (the amazing variable star Dschubba), Delta Centauri, Mu Centauri, Beta Scorpii, Nu Scorpii, Eta Centauri, Gamma Lupi, Lambda Lupi, and 48 Librae.

The Lore of Antares

As one of the two ruddiest of the 1st-magnitude stars, and a member of the zodiac's brightest constellation, Antares has been famed in lore since earliest times. Among its early titles in Mesopotamia were "the Vermillion Star," "the Lord of the Seed," and "the Day-Heaven-Bird." It was also known as Lugal Tudda, "Lusty King," and Dar Lugal, which the Jobeses say was "the King who was lord of lightning." The Jobeses also write of Antares being a symbol of Selk, an Egyptian "reptile goddess or earth mother, who was proclaimed at the autumnal equinox, the season of decline for both earth and the star.'" In ancient China, Antares was, among other things, Who Sing, "the Fire Star" or "Great Fire." The Sogdians called Antares "Maghan sadwis," that is, "the Great One saffron-colored." The Khorasmians called it Dharind, "the Seizer." The Copts knew it as Kharthian, "the Heart." The Jobeses mention that Antares was known in Central Asia as "the Grave Digger of Caravans" because as long as journeyers saw it rise in early morning, they knew robbers and death would stalk their path. Even classical Greece and Rome had interesting alternative titles for Antares. Guy Ottewell comments, "There are dubious but pleasing possibilities that it has been known as Vespertilio—Bat Star, Rustle of Evening—and Insidiata—Star in Ambush." The Cherokee, Pawnee, and other Native American tribes thought that souls at death traveled southward along the Milky Way, trying to survive dangers (including treacherous animals and raging torrents) to reach "the most southerly station" where they could rest for eternity on Antares, the Spirit Star.

The lore and legends of Antares are not just things of the past. There are many references to the star in modern science-fiction. Most memorable of all of these, no doubt, is the song that Lt. Uhura sings in several episodes of the original *Star Trek* TV series. Accompanying herself on a harplike instrument, she sings beautifully a haunting romantic song called "Beyond Antares."

24 ▸ POLLUX

This chapter is written in praise of Pollux. That star has been undervalued by astronomers for a long time. Yet it is the distinctly brighter of the two stars whose paired presence has led to their constellation being called Gemini, the Twins—one of the few brightest and remarkable constellations of the zodiac. Pollux is itself colorful, is close enough to the ecliptic to have dramatic close conjunctions with the Moon and planets, is the closest "giant" star to Earth, and, as recently as 2006, became the first very bright star for which there is truly strong evidence of being accompanied by a planet.

And yet, as we'll see, it will take discussion of these and other wonderful distinctions of Pollux to help give the star precedence, or at least equality, with its dimmer but much more popular "twin."

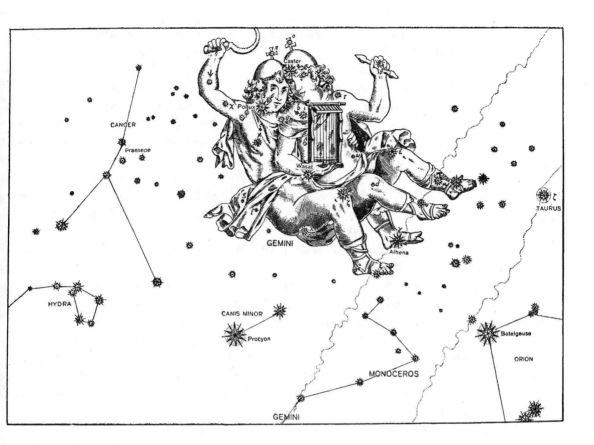

The Less Acclaimed of the Twins

Who are the most famous of all twins? Remarkably, they are not historical people. They are characters in Greek mythology. And their names have been given to the most famous of all pairings of bright stars in our sky: Pollux and Castor.

But if you are a reader of mythology, you might be immediately jolted by something I just did: I gave the name of Pollux first. When discussing the mythological figures it is customary to say "Castor and Pollux." A person would be no more likely to say "Clark and Lewis" or "Costello and Abbott" than "Pollux and Castor."

Yet in astronomy it is possible to speak for quite a while about one of these stars without mentioning the other and, when the two *are* mentioned together, wouldn't it make sense to say first the name of the star that appears brighter? That star is Pollux. The stars Pollux and Castor are not really twins

Gemini, featuring Pollux and Castor.

in brightness, for Pollux is noticeably brighter even to a casual view. More notably, the dividing line between first magnitude and second magnitude comes at 1.50 and this falls between the brightnesses of Pollux and Castor—so Pollux gets its own chapter in this book, while Castor doesn't.

Pollux's magnitude of 1.16 edges out Fomalhaut by only 0.01 magnitude for the title of seventeenth-brightest star in the heavens. But Pollux glows 0.42 magnitude brighter than the combined radiance of the Castor system.

Castor is not even quite the brightest of the 2nd-magnitude stars, by the way. Its 1.58-magnitude light is outshined by 1.50-magnitude Adhara (Epsilon Canis Majoris)—though the latter is overlooked in favor of Sirius in its constellation and is far enough south to be always a little dimmed by atmospheric extinction for observers at midnorthern (and of course higher northern) latitudes.

But guess what? Four hundred years ago, when the Bayer letters were being assigned, Castor got designated Alpha Geminorum. Pollux had to settle for being Beta Geminorum. There are many departures from the letter of the law (letters in order of brightness) in the Greek lettering of the stars that was done by Johannes Bayer and others. But in this particular case, the reason for Castor getting precedence in the lettering was probably the tradition of calling the two mythological brothers "Castor and Pollux"—Castor first, Pollux second.

Now you would think as astronomy developed as a science, learning more about the stars, and especially as amateur observational astronomy became established as a discrete hobby or pastime, the greater brightness of Pollux would overcome the old order-of-naming bias and make Pollux at least as popular an object of talk and attention as Castor.

But that's not so. The reason is that Pollux seems to be a single star but telescopic observation shows Castor to be a spectacular double star—a magnitude 1.93 and 2.97 pair now 4½" apart (and slowly widening). In fact, if you know where to look, there is a third, distant visual companion in the Castor system. And spectroscopic studies have revealed that all three of these stars of Castor are themselves close doubles—a six-star stellar system!

No wonder then that Castor has attracted more interest among amateur astronomers than Pollux.

The Planet of Pollux

But move over, Castor. Pollux is apparently not alone after all. In 2006, researchers presented strong evidence that the single star Pollux possesses a planet—the first such clear identification of a planet for one of the night's brightest stars. Robert Naeye wrote beautifully about this discovery in *Sky & Telescope*. As he noted, astronomers had previously identified about two hundred extrasolar planets but most of them orbiting "obscure stars with quite forgettable names"—or, indeed, usually not a name but a catalog designation. Few of these stars had enough apparent brightness to detect without optical aid. "Finally," wrote Naeye, "a planet has been found around a bright, familiar star dear to the hearts of amateurs worldwide: Pollux." Unlike the planets speculated to circle Vega and Fomalhaut, this one is not just a possible interpretation to explain concentrations of dust in a circumstellar dust disk. In this case, the evidence is Doppler data from the spectrum of Pollux that probably doesn't admit of other interpretation.

What is the planet of Pollux like? Well, we wouldn't expect life on it. It is believed to have at least 2.9 times the mass of Jupiter and take 590 days to execute an almost perfectly circular orbit very roughly 1.6 AU from the star. That's similar to the distance of Mars from our Sun, but Pollux burns about thirty-two times brighter than the Sun. This world must be a terribly hot place. But of course the existence of one planet greatly increases the possibility that other, farther-out, cooler planets may also exist in the Pollux system. A planet of Pollux in the "comfort zone" for Earth life would be about 5.6 AU from the star (farther out than Jupiter in our solar system).

It was actually two groups who made the discovery of Pollux's planet. A member of one of those groups, Geoff Marcy, puts the finding of this world into perspective. "What a wonderful gift," he says, "that anyone, even in the center of a city, can gaze up and see a star that has a planet."

Conjunctions, Occultations, and Color of Pollux

Pollux and Castor are profoundly different stars, both in some aspects of their appearance and in their nature as suns. In several of these categories, Pollux is surely the more interesting.

For one thing, Pollux is considerably closer to the ecliptic than Castor and thus has closer, more impressive conjunctions with the Moon and planets. Actually, after almost every close encounter of the Moon or a planet with Pollux, there comes another superb sight: a more or less compact straight line of the Moon or planet with Pollux *and* Castor. Sometimes the line is shorter than other times. Sometimes it is perfectly spaced: $4\frac{1}{2}°$ from Castor to Pollux, $4\frac{1}{2}°$ more to the Moon or Venus or some other planet. When we compare Pollux and Castor to the two pairings of stars that are much brighter and a little closer together—Alpha Centauri/Beta Centauri and Alpha Crucis/Beta Crucis—we need to remember that those other pairings are very far from the zodiac and never have visiting planets or a Moon with which to form spectacular patterns.

What very few skywatchers know is that Pollux was formerly capable of something that now may only be said of Aldebaran, Regulus, Spica, and Antares, among the 1st-magnitude stars: Pollux could be occulted by the Moon. Nor were such events only possible tens of thousands of years ago. The last occultation of Pollux by the Moon visible from anywhere on Earth took place in historic times—117 BC.

Pollux is also more interesting than Castor in color. Castor and its brightest components are whitish. Pollux has an orange hue that most people can even detect, at least slightly, with the naked eye—perhaps especially because the white Castor is nearby for comparison. The B-V index of Pollux is +1.0, which places it about midway in hue between yellow Capella and more deeply orange Arcturus.

The Nearest Giant

The color of Pollux indicates that it is a K class star. Among the 1st-magnitude stars, Pollux is the closest K star—at 34 light-years from Earth, a little closer than Arcturus. But even more notable perhaps is that Pollux is the closest giant star—meaning the closest star that has swelled and brightened and left the main sequence of stars that contains the Sun, Alpha Centauri, Sirius, Procyon, Altair, Vega, and Fomalhaut. Pollux is still in an early stage of gianthood, though. Its spectral designation is K0IIIb—the "b" indicating the fainter division of the luminosity class III stars, which are "giant" stars less luminous and smaller than the class I "supergiants" and class II "bright giants."

Pollux is thought to be about 8.8 times wider than the Sun, though perhaps only about 1.7 times as massive. It is 32 times more luminous than the Sun—compared to Castor's six stars that, at 52 light-years from Earth have, by odd coincidence, a combined luminosity 52 times that of the Sun.

There is evidence that Pollux may rotate in thirty-eight days—not greatly longer a period than our Sun's. Pollux also seems to have a corona that may resemble that of the Sun.

In championing Pollux as an interesting sun, I don't want to deny that the physical aspects of the Castor system are truly remarkable. Consider the two close-together pairs of stars that one sees as two nearly-touching points of light in an amateur telescope. They take about 445 years to orbit each

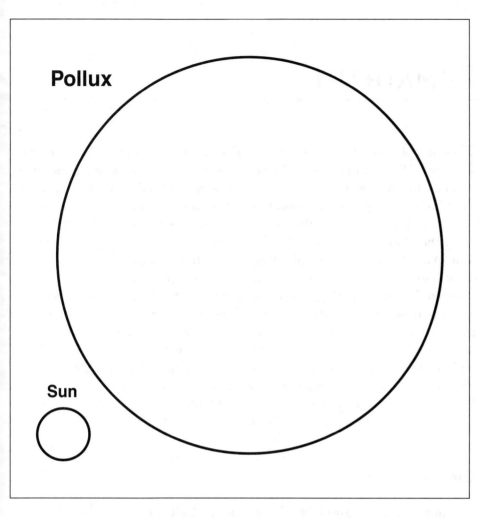

Pollux

Sun

Relative sizes
of Pollux and
the Sun.

other (the separation between them is opening up nicely in the early twenty-first century). The brighter member of the Castor A pair is a star quite similar to Sirius in mass, diameter, luminosity, and spectral type. The brighter member of the B pair is similar to Fomalhaut. By the way, the point of light that the naked eye sees as Castor was probably first resolved into two stars by Giovanni Domenico Cassini (famous for discovering the Cassini Division between Saturn's A ring and B ring). Castor A and B were the first objects beyond the solar system for which there was evidence of being gravitationally bound together in an orbit around each other.

Pollux is only $4^1/2°$ from Castor in the sky. But how far apart are they in space? About 18 light-years. A life form on a Pollux planet would see Castor as a star of almost magnitude –1. But you may be surprised to hear that

25 FOMALHAUT

Few people would associate the star Fomalhaut with a much brighter star very far from it in the sky—Vega. Yet there are several major similarities between the two. First, Fomalhaut lies at almost exactly the same distance from Earth as Vega does. Second, while Vega is the bright zenith star for the most populous latitude in the Northern Hemisphere (40°N), Fomalhaut is easily the closest there is to a bright zenith star for the most populous latitude in the Southern Hemisphere (generally held to be 35°S). Third, Fomalhaut and Vega were the two bright stars discovered in the early 1980s to have a disk of dust circling at a considerable distance from themselves—possibly the dust components of comets. (The twenty-first century brought an exciting follow-up to those discoveries, which I'll get to later in this chapter.)

Why do I start this chapter profiling the connections between Fomalhaut and Vega? Perhaps to counter the fact that Fomalhaut is in many ways the most secluded and uncompanionable of the very bright stars. It stands out from its location in the heavens partly because there is no competition—almost not even any company—for it.

The Lonely Star

Fomalhaut is unquestionably the loneliest of all the 1st-magnitude stars. Earlier in this book, we saw that the much more southerly star Achernar was also secluded—in fact, the 1st-magnitude star closest to it was a full 39° away. And that other star was Fomalhaut. But Achernar, we learned, also has blazing Canopus $39^1/_2$° away from it (in almost the opposite direction from Fomalhaut), other 1st-magnitude stars moderately farther distances away from it, and the Magellanic Clouds forming a fairly compact triangle with it. Also, although in almost any populated land where Achernar can be seen well, Fomalhaut can, too, the opposite is far from true. In fact, at midnorthern latitudes where most of the world's people live, only Fomalhaut, not Achernar, can be seen. Achernar has Fomalhaut and stars like Canopus as seen from southerly lands, but Fomalhaut doesn't have Achernar as seen from northerly lands.

Fomalhaut lies 59° from Altair, 83° from Antares, 91° from Vega, and 93° from Aldebaran. More amazingly, Fomalhaut is the only 1st-magnitude star visible from midnorthern latitudes across a span of about 8 hours of right ascension (Deneb to Aldebaran)—one-third of the way around the entire heavens! Even in the Southern Hemisphere, the span is from Deneb to Achernar—5 hours of RA.

Fortunately, as we saw earlier in the book, Capella, the most northerly of all the brightest stars, comes up so early around latitude 40° N that it actually rises around the same time as Fomalhaut (even though Capella is about 6 hours of RA east of Fomalhaut). But, especially after their co-rising, very few people turn their head far enough to take in both Fomalhaut and Capella in a scan. Capella rises and, for many hours, stays very high. At 40° N, Capella spends about twenty hours above the horizon, passing north of the zenith. At that latitude, Fomalhaut is above the horizon for little more than eight hours. Since its declination is about –30°, Fomalhaut never climbs higher than 20° above the horizon. At that angular altitude, atmospheric absorption on a typical clear night dims an object by almost half a magnitude. So people at 40° N never see Fomalhaut appear brighter than about magnitude 1.6—a bit dimmer than 1st magnitude. From England, Fomalhaut never appears brighter than about magnitude 2.2. And observers as far north as southern Alaska and southern Scandinavia never see Fomalhaut come above the horizon at all.

By the way, it's very interesting that Fomalhaut happens to be the only 1st-magnitude star in the large declination range between about –26° (Antares) and –53° (Canopus). This is what makes it the zenith star for most people who live at middle latitudes in the Southern Hemisphere. It's also what makes it what we might call "the border star" or "transition zone star" or "ultimate low star" for most people who live at middle latitudes in the Northern Hemisphere.

Now someone might object that another 1st-magnitude star, Antares, is only 3° farther north than Fomalhaut. That's true. But Antares is in such a bright constellation in such a rich, fascinating overall region of the heavens that few people would think of isolating it to be considered an Australian zenith star or as the ultimate low (very bright) star for northerly lands. Far-south observers think more in terms of having the whole brilliant Scorpius and the dreamy grandeur of the Sagittarius Milky Way overhead. Folks in Canada and the United Kingdom usually think of Scorpius being the one really bright constellation that they can just see most or some of, but dimmed greatly from the true grandeur it has in lands where it gets reasonably high in the sky.

The Autumn Star and Lonely Lighthouse

The loneliness and lowness of Fomalhaut would seem to doom it to being far less often seen or appreciated than most of the brightest stars. It's not in the sky long and is low enough to be hidden by trees or buildings and dimmed by haze and thick atmosphere (in these respects, for Northern Hemisphere viewers, it is opposite of Vega, whatever other surprising similarities the two may have). But when loneliness and lowness become extreme enough they can, paradoxically, start working in favor of getting a star seen and esteemed. After all, if Fomalhaut is the only fairly bright point of light in a vast area of heavens, even a casual observer may notice it. Or, a novice amateur astronomer, learning that Fomalhaut is the only show in town if one wants to see a bright star in the south on autumn evenings, will make a special point of looking for Fomalhaut. The star's lowness may also put it right at eye level for people who normally don't pause to look *up* at night.

One thing is certain: Fomalhaut deserves to be called "the Autumn Star." Capella might deserve that title also, but it is a star of several seasons. For observers at midnorthern latitudes, autumn is the only season for Fomalhaut. Unless you stay up very late at night in late summer or go out right at nightfall in early winter. (About the latter time and season, I must admit: I'll never forget that it was at nightfall near Fomalhaut, low in the southwest sky down our country road, that I saw my first comet—dimly with the naked eye and with a nice streak of tail in finderscope—back in January 1970.)

Fomalhaut is near the southern edge of that vast region of the autumn evening sky filled with huge, dim water-related constellations. The region, which spans the lower half of the sky from southeast to southwest, is usually called "the Water" or, on rare occasions, "the Great Celestial Sea." I sometimes think of Fomalhaut as a lonely lighthouse, or lighthouse beacon, near the southern shore of that heavenly ocean.

People at midnorthern latitudes may first feel it is a tremendous contrast to have such a great expanse of dimness and darkness in the night when days are fired up with the hurry of migrating birds, warm glow of pumpkins, and especially the bright changing colors of the leaves. But once the birds are gone, the pumpkins harvested, and the leaves fallen and browned, the sparsely and faintly starred vast of the Water and its solitary lighthouse Fomalhaut seem very appropriate.

By early November for many of us, the dark southern sky is consonant with our melancholy feelings about the passing of the last gleam of the living world's growth and activity for the year. Still, we know it is all part of the yearly cycle, that green will flame and flowers grow again in spring. This helps us actually appreciate the somber mood, relish the melancholy.

Fomalhaut's Company

Fomalhaut may not have any close neighbors for us to readily observe it with as it crosses the sky. But farther afield in the sky, there are several stars that have long-distance visual connections to it.

Two such stars are those that form the west side of the famous Great Square of Pegasus. These stars are oriented in a nearly perfect north–south line. Trace the line from 2.5-magnitude Beta Pegasi through 2.5-magnitude Markab and then extend it 45°—about 3½ times its own length—and it brings you right to Fomalhaut. An interesting point is that the west side of the Great Square and the extended line from it to Fomalhaut almost exactly coincides with the 23h line of right ascension. It again seems appropriate to the loneliness of Fomalhaut that it lies virtually on the line that begins the very last hour of RA, highest on the meridian at the start of the very last hour of sidereal time.

A line extended south from the two stars of the *east side* of the Great Square of Pegasus can be used to pass fairly near one of the few bright 2nd-magnitude stars among the Water constellations. The distance is a little more than twice the distance between Alpha Andomedae and Gamma Pegasi (east side of the Great Square), a total of a little more than 30 degrees. The star pointed to (or nearly pointed to) is Diphda, also known as Deneb Kaitos.

The name Deneb Kaitos means "tail of the whale," the whale being the constellation Cetus. Although Deneb Kaitos is designated Beta Ceti, it is actually the brightest star in Cetus, at magnitude 2.0. On autumn evenings, it trails Fomalhaut across the south sky, a dimmer follower than the 1st-magnitude star. At times, Deneb Kaitos is at the same altitude as Fomalhaut but it really lies 12° of declination farther north than Fomalhaut. The distance between the two stars is 27°—a fairly large gap but in this otherwise dim expanse of heavens small enough to make Deneb Kaitos a sort of "closest" major acquaintance of Fomalhaut. As a matter of fact, the visual relationship between these two stars has been commemorated by star myth—and leads us naturally into a discussion of the legends associated with Fomalhaut and with some of the names of both it and Deneb Kaitos.

The Name Fomalhaut and the First Frog

Let's start with the now-widely accepted name of our featured star—Fomalhaut. Many people who first encounter this name are puzzled as to how to pronounce it. Some, having some familiarity with French, suppose that the *aut* at the end of the name should be pronounced like the English word *owe*—thus *FOHM-uh-loh*. In reality, the name is from the Arabic *fam al-hut*, "the mouth of the fish" (the constellation Piscis Austrinus, the Southern Fish). The correct way to pronounce the name today is *FOHM-uh-lawt*.

R. H. Allen has said of Fomalhaut that "[n]o other star seems to have had so varied an orthography." Among the spellings of the name he reports as having been used (from the Middle Ages to the nineteenth century) are: Fomahant, Fumahaud, Fomahandt, Fumahant, Fumahaut, Fumalhaut, Phom Ahut, Fomahand, Fontabant, Fomauth, Phomaut, Phomault, Phomant, Phomaant, Phomhaut, Phomelhaut, Phomalhaut, Fumalhant, Fomahaut, Phomahant, Fomahout, Pham Al Hutand, Fomalhani, and Fomalcuti! Sir William Herschel used yet another spelling when writing to his sister Caroline: "Lina—Last night I 'popt' upon a comet. . . . between Fomalhout and Beta Ceti."

Fomalhaut has often (though not always) been pictured in the mouth of Piscis Austrinus, often a mouth that is seen as catching the sparkling (star-speckled) stream of water being poured out by the constellation just to the north of Piscis Austrinus—Aquarius, the Water-Bearer.

The Arabs of the Middle Ages did not only imagine Fomalhaut as the mouth of the ancient Greek fish. And here is where the connection of Fomalhaut and Deneb Kaitos comes. Or rather, Fomalhaut and Deneb Kaitos by its other name, Diphda. The two stars were known to the early Arabs as Al Difdi al Awwal (Fomalhaut) and Al Difdi al Thani (Diphda). The names respectively mean "The First Frog" and "The Second Frog." Fomalhaut is the first probably because it leads Diphda on the nightly journey across the sky.

There are some unsettling but perhaps unsubstantiated associations of Fomalhaut with ancient monsters and monstrous gods. Robert Burnham Jr. mentions an association of Fomalhaut with the most terrifying of the Titans, Typhon, whom Zeus and the gods managed to imprison beneath Mount Etna—explaining the mountain's occasional volcanic activity. Burnham also alludes to an identification of Fomalhaut with the savage fish-god Dagon, best known for having his temple torn down by Samson in the Bible.

Camille Flammarion was a flamboyant French astronomer who popularized astronomy in the late nineteenth century. He claimed that Fomalhaut was "Hastorang" in Persia—in 3000 BC, when it was considered one of the four Royal Stars, the four Guardians of Heaven. (The other Royal Stars were Regulus, Aldebaran, and Antares.) Around 500 BC Fomalhaut was also supposed to be the object of sunrise worship in the temple of Demeter at Eleusis.

Scholars have often argued that Fomalhaut was, with Achernar and Canopus, the "Tre Facelle" mentioned by Dante in *The Divine Comedy*. Allen writes that sixty years before the time of his (own) classic book on star names (therefore around 1839), someone called Boguslawski thought that Fomalhaut might be "the Central Sun of the Universe."

Fomalhaut as a Sun

Boguslawski notwithstanding, Fomalhaut is actually a rather modest sun. Of the twenty-one stars of greatest apparent brightness, only a few are smaller and less luminous. Since Vega is at essentially the same distance from us as Fomalhaut and appears more than a magnitude brighter, we know the true brilliance of Vega must be a magnitude brighter, too. It may come as a surprise then to learn that Fomalhaut is a spectral class A3 compared to Vega's A0 and, most important, that Fomalhaut's mass is thought to be only 0.2 solar mass less than Vega's—2.3 solar masses to Vega's 2.5. But a few spectral subclasses away and even a seeming slight difference in mass can make a big difference in the color and speed of evolution of a star. Vega is decidedly blue-white and Fomalhaut very definitely white. The big difference in their luminosity is not due to Vega's being somewhat hotter (therefore bluer), though. It is that Vega has a lot more surface area—its average diameter is about 2.5 times our Sun's, whereas Fomalhaut's is only about 1.7 times our Sun's.

By the way, Fomalhaut is one of those stars whose color some observers of the past called red. But this was no doubt because they saw it low and reddened by the thick air and haze low in the sky. The surface temperature of Fomalhaut is about 8,500 K—that compares to 8,400 K for Deneb, which everyone agrees is white. Blue-white Vega has a B-V color index of 0.00; while white Deneb is 0.09; very, very slightly yellow Altair is 0.22; slightly yellow Procyon is 0.43; plainly yellow Capella, 0.80. The B-V color index of Fomalhaut is 0.14.

Fomalhaut is perhaps only about 200 million years old, but will probably not live to be more than about 1 billion years old before becoming a white dwarf star and slowly fading away. And, as is the case with Vega, the 1 billion years may not be enough time for life to develop on any planets orbiting Fomalhaut. Also, as in the case of Vega, we actually have evidence that there may be at least one planet orbiting Fomalhaut.

Fomalhaut's "Kuiper Belt" and Possible Planet

Back in the early 1980s, the satellite known as IRAS detected in infrared the presence of a large circumstellar disk of dust around Vega—and also around

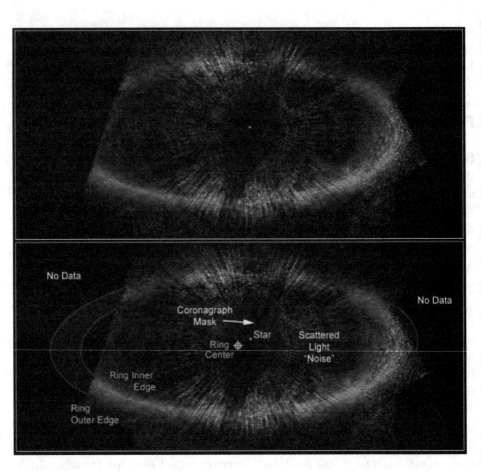

Labels in image: No Data, No Data, Coronagraph Mask, Star, Scattered Light "Noise", Ring Center, Ring Inner Edge, Ring Outer Edge

Hubble Space Telescope images of the dust disk around Fomalhaut.

Fomalhaut. In 2002, studies of the vicinity of Vega discovered clumps in its disks, structures presumed to be caused by two planets. Thus Vega became the first of the very bright stars for which there was some evidence of having planets. But Vega's exclusive status in this respect didn't last long.

In 2004 Hubble Space Telescope imaged Fomalhaut's "debris disk" in far greater detail than had ever before been achieved. Much more was learned about it. The estimated mass of the material in the disk turns out to be 50 to 100 Earth-masses. The fact that the disk is located 133 to 158 AU from the star suggests that the disk could be analogous to our own solar system's Kuiper Belt, which includes Pluto and large numbers of similar—if mostly much smaller—objects. It's true that the Kuiper Belt lies only about 30 to 50 AU out from our Sun but it's not unreasonable to think a star several times as massive as the Sun would have its own Kuiper Belt farther out. It's also true that Neptune gives a sharp inner edge to the Kuiper Belt. And that's what makes it so interesting that the Hubble imaging reveals Fomalhaut's disk to have sharp inner and outer edges. What's more, the disk is shown to be lopsided—offset by 15 AU. The most likely cause is the presence of one or more massive planets at 50 to 70 AU out from Fomalhaut.

Fomalhaut's Fellow Wayfarer

Fomalhaut is lonely in our sky, but how lonely is it in space? At least seven star systems are within 10 light-years of it but none of them is very luminous. Not until you travel 15.7 light-years from Fomalhaut do you come to a star with a Greek-letter designation, Epsilon Indi—which is magnitude 4½ from Earth but closer to us than to Fomalhaut. (By the way, Epsilon Indi is considerably cooler and less luminous than our Sun but has always been considered one of the prime candidates among our neighbor systems in the search for extraterrestrial life.)

How far in space from Fomalhaut are brighter, better-known stars? Altair is 21.9, Alpha Centauri 24.8, Sol 25.2, Sirius 28.3, Procyon 33.2, and Vega 36.1 light-years from Fomalhaut.

Again like Vega, Fomalhaut seems to be a single star. But Fomalhaut has a strange circumstance. Another star is going through space with it, at a distance of 0.9 light-years. That's too far for there to be a real gravitational connection. But as Burnham writes, "We might speculate that Fomalhaut and its distant companion are the last surviving members of a low density cluster which gradually dispersed and scattered, as such clusters as the Coma Berenices group appear to be doing today." Remarkably, Fomalhaut, Vega, and Castor have similar enough motions through space to raise the possibility that the three are members of what was once a stellar association.

The star less than a light-year from Fomalhaut is GC 31978, though is perhaps best known by its variable star designation: TW Piscis Austrini. It is a flare star, although its brightness variations seem to be from only about apparent magnitude 6.44 to 6.49 over a period of ten days as seen from Earth. TW is estimated to be very slightly closer to us than Fomalhaut—24.9 light-years, with a margin of error of plus or minus 0.2 light-years. The star is estimated to be 0.8 the mass and maybe 0.8 the diameter of the Sun. Its luminosity is only about 0.12 that of the Sun. We note TW almost 2° south of Fomalhaut in our sky.

But in Fomalhaut's sky, TW Piscis Austrini would be a beautiful orange star of magnitude –0.5. And what about from the opposite perspective? As seen from TW, Fomalhaut would shine at about magnitude –6, several times brighter than Venus ever gets in Earth's sky.

26 BETA CRUCIS

Beta Crucis is almost a half-magnitude dimmer than Alpha Crucis but is tied with Deneb for the rank of nineteenth brightest of night's stars. It lies only about $4^1/_4°$ from Alpha Crucis and marks the eastern end of the transverse beam of the 6°-tall Crux the Cross. In that position, it is a little closer than any of Crux's main stars to Alpha and Beta Centauri.

All of the preceding is meant to emphasize the fact that Beta Crucis's greatest attraction is its being a part of the glorious Centaurus–Crux–Carina section of far-south Milky Way. But one could also argue that the much closer vicinity of Beta Crucis—say the 1° to 2° field you could fit into the view of most telescopes—is more impressive than that of any of the other 1st-magnitude (and brighter) stars in this region of the sky. Furthermore, the star holds a few distinctions that make it interesting in and of itself. For instance, no other 1st-magnitude star has a companion whose nature is so uncertain to us. And one of the proper names attached to Beta Crucis is not only of unknown origin but has a meaning whose appropriateness no scholar has been able to explain.

Beta Crucis and the Jewel Box

Beta Crucis lies little more than a degree from the northern edge of the most impressive of all naked-eye dark nebulae, the Coalsack. The Coalsack measures 7° by 5°, thus comparable in size to the Southern Cross it lies next to. Steve O'Meara has therefore called it "the Shadow of the Cross." But an older fancy is that this is a hole in the sky left when the Southern Cross was lifted out of it and, apparently, moved just to the side. Science now tells us that the Coalsack is, in reality, two dark, overlapping nebulae that are 610 and 790 light-years distant—thus a little less than two and about two and a half times farther than Alpha, Beta, and Delta Crucis (the main star of Gamma Crucis is only 88 light-years from Earth).

Now Beta Crucis may be very close to the north of the Coalsack but Alpha Crucis is even closer to—within a degree of—the dark splotch's western edge. So when I said above that the telescopic vicinity of Beta Crucis was more impressive than that of the other bright stars in this region (once you get beyond the splendid companions of Alpha Centauri A and Alpha-1 Crucis), what did I mean? As any avid observational astronomer who lives in or has observed in the Southern Hemisphere realizes, I mean NGC 4755, better known as the Kappa Crucis Cluster, or just "the Jewel Box."

This object, one of the richest and one of the few finest of all open clusters, is a glittering treasure of almost three hundred stars in a 10' wide area. At the cluster's estimated distance of 4,900 light-years, the angular size would work out to a real size of only 14 light-years.

This cluster sparkles just 1° southeast of Beta Crucis. Steve O'Meara points out that the latter-day claims that Kappa Crucis is a magnitude 7.2 star at the cluster's center, or a magnitude 5.9 star in the cluster, are wrong. The designation Kappa Crucis, he maintains, was given by early explorers to the 4.2-magnitude naked-eye fuzzy spot of radiance that is the entire cluster. The first slightly more detailed study of the cluster was made by Nicholas Louis de Lacaille in 1751–53. But the name "the Jewel Box" has been derived from Sir John Herschel's enraptured nineteenth-century description of the cluster and the varied gemlike colors of its members when viewed through a telescope large enough to reveal the hues "which give it the effect of a superb piece of fancy jewelry."

Whatever you choose to call it, the Jewel Box forms a superb accompaniment to Beta Crucis and, with its brightness, an amazing contrast to the Coalsack, whose northern edge the cluster is so near.

Becrux and the Mystery of Mimosa

Just as the Greek-letter designation Alpha Crucis, or rather Alpha of Crux, was contracted into Acrux, Beta Crucis has been called Becrux. I am not aware, however, who first used Becrux. It seems to have been in the twentieth century because, whereas R. H. Allen, writing in 1899, mentions Acrux, he does not refer to Becrux (oddly, whereas Allen briefly discusses several of the other stars of Crux, ones known only by their Greek-letter designation— Gamma Crucis and Kappa Crucis—he makes no reference whatsoever to Beta Crucis). Neither Paul Kunitzsch nor Guy Ottewell, both writing in the late twentieth century, bother to mention the name Becrux, though they do mention Gacrux (pronounced *GAH-kruks*) for Gamma Crucis. The reason for this is probably because, unlike Alpha Crucis and Gamma Crucis, Beta Crucis does already have what seems a proper name that is "real" (no mere manufactured contraction). The name is Mimosa.

The problem is that neither Kunitzsch nor Ottewell—nor perhaps anyone else—knows who coined the name Mimosa for Beta Crucis or why it was thought appropriate for the star. The word *mimosa* is used for a genus of plants, some of which shrink from the touch. Kunitzsch points out the word is from the Latin *mimus*, "an actor" (and literary scholars speak of *mimesis* in literature). Ottewell chooses to go back to the Greek *mimos*, which he translates as "mime." But in what way is Beta Crucis like a mime or actor—or a kind of plant? (By the way, the name could not refer to color because mimosas are yellow and Beta Crucis is blue-white.)

Beta Crucis as Suns

Like other members of the Scorpius-Centaurus Association of stars, Beta Crucis is very hot and very blue. The spectral type is B0.5 and the surface temperature is about 27,500 K, making it just about as hot as any 1st-magnitude star. What's more, Beta Crucis is classified as being in luminosity class III—a giant—although James Kaler says this may be jumping the gun. The hot star produces a lot of ultraviolet radiation. Although visually its luminosity is about 3,000 times that of the Sun, if all radiation is added in, Beta Crucis turns out to be about 34,000 times more luminous than the Sun.

Calculations suggest that the diameter of Beta Crucis should be 8.1 solar diameters, but actual measurements indicate it is 8.4. Yet, as with Alpha Crucis and Beta Centauri, there is also a close companion in this system. In this case, however, the evidence suggests that the close companion is not very luminous. The two stars revolve around a common center of gravity in almost exactly five years and may be separated by about 8 AU. The primary star would have fourteen times the mass of the Sun and be fated to become a supernova.

There is an 11th-magnitude star about 44" from Beta Crucis but it is probably just an optical, not a physical companion.

Beta Crucis is a variable star—but which kind? Robert Burnham Jr. says that it is a Beta Canis Majoris star that ranges by just 0.06 magnitudes with a period of 5 hours, 40.5 minutes. But Kaler (writing a few decades later) calls it a "Beta Cephei star," a "multiply-periodic" variable star whose magnitude ranges between 1.23 and 1.31 in periods of 5.68, 3.87, and 2.91 hours.

Beta Crucis has only about half the metal content (half the amount of elements heavier than hydrogen and helium) that our Sun does. This would suggest that it is a particularly young star. Its age may be about 10 million years.

DENEB

27

A particular stellar sight is visible from midnorthern latitudes. The time of night that this sight is visible depends upon the time of year, of course. But the most often witnessed combination is evening during September or October.

At that time, a bright—though not blazingly bright—star shines nearly overhead. It may seem like little more than a pleasant, gentle anticlimax to brilliant Vega's zenith-passing of two hours earlier. This less radiant-looking follower of Vega is also part of a rather bright constellation that it cannot dominate overwhelmingly the way Vega and Altair do their dimmer constellations Lyra and Aquila. In fact, if your location is far from city lights and the

The Summer Triangle. The three brightest stars in the picture (in order of brightness) are Vega at the top, Altair to the right, and Deneb to the left.

night is moonless and clear, the background glow of the Milky Way's Cygnus Star Cloud may almost drown out—well, at least distract you from—the major stars of Cygnus, including this modestly brightest star of the Swan.

The star, however, is Deneb. And gentle though its light may seem, there is a truth modern science has revealed about it: Deneb, dimmest star of the Summer Triangle in our skies, is in reality a sun much mightier in luminosity than any other within several thousand light-years of our Earth.

The Tail of the Swan

There's no doubt that Deneb represents the tail, not the head, of Cygnus, the Swan. The name Deneb itself is from the Arabic *Dhanab*, which means "tail."

Elsewhere in the heavens, we see a number of other stars with Deneb in their name: Denebola (tail of Leo, the Lion), Deneb Kaitos (tail of Cetus, the Whale), Deneb Algiedi (tail of Capricornus, the Sea-Goat)—and several stars whose Deneb-including names are rarely used. But Deneb is *the* Deneb—the most important star marking the tail of a constellation figure, the beautiful swan. Or at least that's the way things stand today: Deneb was known to the later medieval Arabs as *dhanab al-dajaja*—"the tail of the Hen." Most of us today much prefer to think of this lovely star as being the tail of a graceful swan.

There's certainly reason to compare the main pattern of Cygnus to the form of a swan. The line from Gamma Cygni to Eta Cygni and Beta Cygni, or just the line from Eta to Beta, is much longer than the line from Gamma to Deneb. This long line well represents a swan's long neck. Likewise, it represents the longer part of the upright beam of the "Northern Cross" asterism—the part before the transverse beam (in the imagining of the Swan, that transverse beam is the wings of the bird). And this means that Deneb marks the top of the imagined Northern Cross. Deneb can be a guide to many of the great sights in Cygnus (including nearby double stars 61 Cygni, Omicron-1 and -2 Cygni) and nebulae (including the North America Nebula, which may or may not be lit by Deneb).

No observer at midnorthern latitudes should fail to notice what happens to the orientation of the Northern Cross as it heads down toward the northwest horizon. The Northern Cross, when it reaches the horizon, at last stands upright—with Deneb the beacon shining at its top.

Although in the dark, natural skies of ancient times, Cygnus may have been almost overwhelmed by the Cygnus Milky Way, it's still surprising to me how little lore about Cygnus and Deneb seems to have come down to us. The

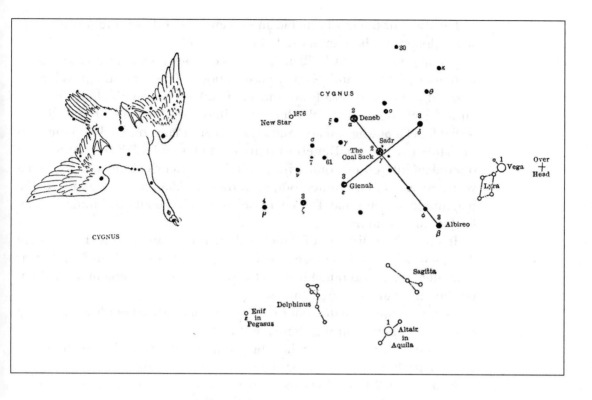

most famous swan in Greek mythology is, of course, the swan that Zeus turned into to ravish Leda.

More than a thousand years ago, before they pictured a hen in the stars of Cygnus, the Arabs called Deneb by the title *al-ridf*. Paul Kunitzsch translates this as "the One Sitting Behind the Rider (on the same animal)" or just "the Follower." R. H. Allen translates it as meaning "the Hindmost." Kunitzsch notes that the title may be in reference to Deneb's relation to Delta, Gamma, Epsilon, and Zeta Cygni, which were called *al-fawaris*, "the Riders." The title *al-ridf* led to a later alternate name for Deneb that one still sometimes encounters: Arided.

How Bright, How Far?

Astronomers know that Deneb must have a greater true brightness than any of the other 1st-magnitude stars. Only five of the twenty-one brightest stars are "supergiants" of luminosity class I. The least luminous of the subclasses within this class is Ib—which Canopus belongs to. More luminous are Betelgeuse and Antares, which belong to class Iab. Some authorities assign Rigel to Iab also, but the most reliable and recent source I can find lists it as a Ia star—along with the even more luminous Deneb.

Just how fantastically brilliant Deneb is obviously depends on its distance from us. But that quantity is very poorly known. The minimum possible distance of Deneb is probably around 1,500 light-years. That's almost twice as far as Rigel and gives an absolute magnitude of –7.5, compared to Rigel's –6.6 if Rigel is 800 light-years from us. For years, many sources listed the most probable distance of Deneb as between 1,600 and 1,700 light-years and offered an interesting comparison with Altair. Altair's distance is 16.7 light-years. So, as I often pointed out, even though Altair appears almost half a magnitude brighter than Deneb, Deneb was really about one hundred times farther away from us.

It was also this distance of Deneb—about one hundred times farther than Altair—which I used to make a very telling calculation. If Deneb is that far away, then it emits as much light to the universe on one September night as our Sun does in an entire century.

That's a powerful statement to make, especially while people are standing outside together, admiring Deneb in the sky.

But some of the latest studies suggest that Deneb may be much farther away—and therefore even more monumentally luminous.

Estimates of 2,100 light-years, even as much as 2,600 light-years, are being given for the distance to Deneb. It's quite possible that Deneb is as much as 160,000 times more luminous than our Sun.

In his *Astronomical Companion*, Guy Ottewell tried to put in perspective how rare a star like Deneb is:

> In our 6.5 parsec [21.2 light-year] sphere, Sirius was without rival. But as soon as we make the view wider, stars enter in which are several times as powerful as Sirius; in this sphere there must be many stars . . . on the level of Sirius—Aldebaran's province is packed with Sirius-villages. Enlarging the view farther, we find that we belong in the kingdom of Canopus, a star dominating all the petty Aldebarans for fifty parsecs around. But, at even more enormous distances from us and from each other, there are other Canopuses, and at last the far-off unapproachably brilliant emperor—Deneb.

Then Ottewell takes us on the opposite journey—outward *from* Deneb: "Around Deneb there is . . . a sprinkling of Canopuses, around each of these a court of Aldebarans, around each of these an army of Siriuses, around each of these a band of suns and a thick populace of red dwarfs. *For each Deneb there may be a thousand million stars.*" [Emphasis added.]

James Kaler says that if Deneb is 2,600 light-years away, it would be intrinsically about the brightest star of its temperature and spectral type in the entire galaxy. Of course, the key there is "of its temperature and spectral type." What is the spectral class and temperature of Deneb? What other qualities does it have that set it apart from other stars?

Deneb as a Sun

Deneb's spectral type is A2. The stars of the Summer Triangle are all spectral type A stars—but tremendously different. Deneb is hotter than Altair but cooler than Vega. But whereas type A0 Vega has an average diameter about 2½ times that of the Sun, the diameter of Deneb is roughly 200 times greater than the Sun's. If placed where our Sun is, Deneb would fill the solar system out almost to Earth's orbit. (A slightly lower estimate for Deneb's size is 180 solar diameters—which would be about 100 times that of Altair. So Deneb might be not just 100 times farther but 100 times wider than Altair). Betelgeuse and Antares are 4 or more times wider than Deneb but their surfaces are very cool for a star. Deneb's surface temperature is about 8,400 K.

We read of blue giant stars—like Rigel—and red giant stars—like Betelgeuse. But why is Deneb white? The answer is that we are just catching it

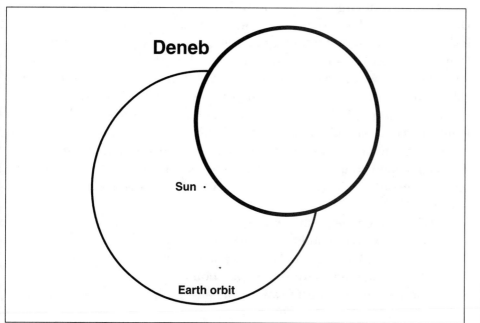

Relative sizes of Deneb and Earth's orbit.

during what might be a rather brief (by stellar standards) time that it is in this spectral type. So massive a star can radically shift spectral type back and forth from O or B to M in maybe as "brief" as tens of thousands of years, as its core keeps getting denser and fusing heavier element after heavier element in an attempt to stave off gravitational collapse and resultant supernova. How far has Deneb progressed in its rapid evolution? We know it has used up the hydrogen in its core. But beyond this, as James Kaler says: "Just what it is doing, we do not know."

Deneb must have begun its short life as a star of about 25 solar masses. Right now, it is constant in its light output but its spectrum is observed to be slightly variable. What's impressive is the amount of mass it is constantly blowing off. According to Kaler, Deneb loses mass at about 100,000 times the rate our Sun does. But it will surely not lose enough mass in its brief lifetime for it to avoid going supernova.

Bound for Deneb

We've already discussed how Vega is the brilliant star most deserving of the title "destination star" because it lies not far from the "Apex of the Sun's Way." The latter, however, is the point in the heavens to which our solar system is heading if we measure our motion relative to a number of our neighborhood stars. If we consider a much larger number, or our place in the rotating galaxy at large, the spot to which we are heading may be closer to Deneb. When we look toward the Large Sagittarius Star Cloud, we are staring in the direction of the center of our galaxy. But the galactic longitude of Deneb is almost 90° and it lies very near to the galactic equator. When we look towards it we are staring forward in our spin around the center of the galaxy. We are seeing the approximate place to which the galaxy is rotating us.

This position of Deneb always reminds me of a seldom-quoted stanza of Henry David Thoreau's. I've always wanted to research which nineteenth century astronomer's ideas about our celestial destination Thoreau must have read about and had in mind when he wrote these lines, based on his journey on the Concord and Merrimack rivers in August and September 1839:

> Who would neglect the least celestial sound,
> Or faintest light that falls on earthly ground,
> If he could know it one day would be found
> That star in Cygnus whither we are bound,
> And pale our sun with heavenly radiance round.

A poem of my own called "The Painter of the Skies" includes some lines that may be interpreted in two ways. But one of the ways to interpret them is to identify the star mentioned as Deneb and the journey as the one we take around the galaxy:

> With one bright star before us set
> And Earthspeed million-star well-met,
> We ride along.

REGULUS

28

Regulus has been called a king by a great number of cultures—and almost as far back as history goes. The ancient Greeks named it Basiliskos, which means "Little King" and which today's leading authority on star-names, Paul Kunitzsch, notes as "having obvious origins among the Sumerians and Babylonians." To the Romans, it was *stella regia*, "the Royal Star." Copernicus called the star Regulus in his landmark *De revolutionibus orbium coelestium* of 1543. But Kunitzsch says the claim that Copernicus was the first to give the star this name is untrue, for it appears elsewhere at least as early as 1522. The current name is the diminutive of the Latin *Rex*—thus, as with the ancient Greeks (and no doubt intended as such by some Renaissance scholar), "the Little King."

The Little King

Why is this star, outshined by twenty others in the heavens, regarded as a king?

Presumably because it is the brightest and most strategically placed of the stars of Leo—a constellation pictured as the lordly lion, king of beasts, in many cultures far back into antiquity. Why were the stars of Leo first and then lastingly called a lion? These stars really do resemble the outline of a crouching or recumbent lion—at least as much as any connect-the-dots stick figure of stars can. Regulus is the heart of the Lion from which curls up the

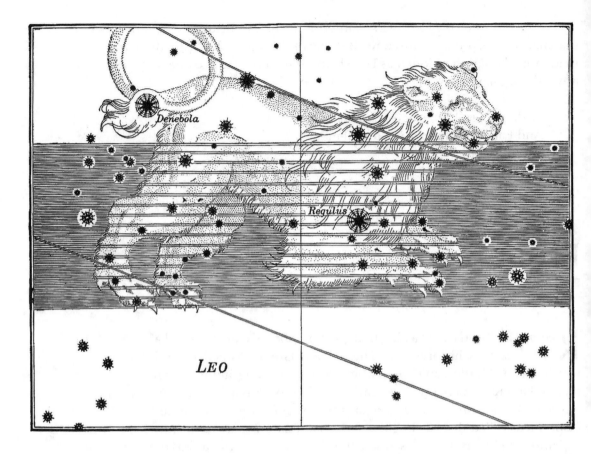

Denebola

Regulus

LEO

Leo and Regulus. 16°-tall hook of stars that resembles the outline of a lion's chest, mane and head. This pattern, including Regulus, forms a backward question mark—Regulus as the dot at the bottom—that is also famous as the asterism called "the Sickle."

Another reason why Regulus and its constellation might have first been seen as a lion is the fact that it was the place of the Sun at the summer solstice in a key period of very ancient history. The constellation and star that marked the summer solstice—the time of the Sun's highest pass across the sky—could only be associated with the animal most like the Sun in fierceness and power. This was true around the middle of the third millennium BC—when the Egyptians noted it along with the heliacal rising of Sirius as the signal that the life-giving flood of the Nile was about to occur.

What, over hundreds and thousands of years, changes the season when certain zodiac constellations contain the Sun, and which star the Earth's spin axis points to as a "north star"? It is the tremendously slow wobble of Earth's axis called "precession"—which completes one circle every 25,800 years.

Discovering the first aspect of this phenomenon—"precession of the equinoxes"—is credited to the Greek astronomer Hipparchus. Around 130 BC, he made the discovery by noticing long-term changes in the positions of two stars—Regulus and Spica. In the case of Regulus, he was comparing the star's position in his time with the location of it mentioned on Babylonian tablets dating all the way back to 2100 BC (almost a thousand years before the most likely time of the Trojan War and Hebrew Exodus).

An aspect of Regulus's position in the heavens that doesn't change much is its proximity to the ecliptic, the Sun's path in the heavens. This proximity intensified further ancient astronomical and astrological regard for Regulus because having Regulus be the very bright star closest to the ecliptic also means that it is the most common bright target for conjunctions with the Moon and planets. I'll have more to say about my own experiences of those glorious conjunctions at the end of this chapter. But let me immediately note that if I and some other investigators are right, then Regulus played a key role in the saga of the Star of Bethlehem—by being the star of kings near the very close conjunctions of Venus and Jupiter in August 3 BC and June 2 BC.

Regulus doesn't change its proximity to the ecliptic much over hundreds or even thousands of years because its proper motion is not exceedingly great. That motion is also almost exactly westward. In the twenty-first century, the Sun passes less than $1/2°$ south of Regulus early in the Universal Time day of August 23 (August 22 in a leap year such as 2008).

Pumpkin, Raindrop, or Bullet

Much of interest about Regulus as a sun has been known for a long time. Astronomers have known that this star has a proper motion companion star, visible through even binoculars at magnitude 7.7 a healthy 3' from Regulus's magnitude 1.35 glow. It is actually a binary companion— magnitudes 8 and 13, separation decreasing from 4" to perhaps less than $2\frac{1}{2}$" since its duplicity was discovered in 1867. In space, this orange dwarf/red dwarf pair are more than twice as far from each other as the Sun is from Pluto, and take about a thousand years to orbit around each other. The orange dwarf is not a feeble star, for it possesses about a half the sun's luminosity and somewhat resembles Alpha Centauri B. From Regulus, the Regulus BC pair would shine at about magnitude –7—for they are about one hundred times farther from Regulus than the Sun is from Pluto, pursuing an orbit around Regulus that takes them at least 130,000 years to complete. From a planet in the B-C system, Regulus would appear several times brighter than a Full Moon on Earth.

Astronomers have long known that Regulus is a hot spectral type B star. In the 1990s, Hipparcos satellite data pegged the distance of Regulus as 77.5 light-years. This makes it about 140 times as luminous as the Sun (240 if ultraviolet light is included) and the closest B class star to us that is a main-sequence star. (By the way, Hipparcos also helps us determine that there are five known star systems within 10 light-years of Regulus, the closest being 6.6 light-years from it, with the stars 40 and 39 Leonis being 13 and 15 light-years, respectively, distant from Regulus.)

Much of interest about Regulus has been known for centuries or at least since last decade. But very recent discoveries about the star have been nothing short of stunning. In a study published in 2005, Harold McAllister and his colleagues at Georgia State University revealed findings made by getting measurements of Regulus with powerful interferometry (a combining of information from an array of linked telescopes). They found that Regulus is not only oblate—that is, its polar diameter smaller than its equatorial diameter—but strongly so. Years ago, the only very bright star known to be quite oblate was the A7 star Altair. Yet Altair's polar diameter is only 14 percent less than its equatorial. For Regulus, this figure is 32 percent. What causes the oblateness is rapid rotation. B stars, like those of the Pleiades, are known to be rapid rotators (it has an effect on their spectral lines). And indeed, much more oblate than even 1st-magnitude B-spectral-type star Regulus is the 1st-magnitude B star Achernar (Achernar's polar diameter is a whopping 56 percent smaller than its bulging equatorial width).

There are consequences of being oblate. In polar measurement, Regulus is 3.15 times wider than our Sun, but in equatorial measure 4.15 times. Being farther from the center of the star means the surface material of Regulus experiences less gravity, and that makes it less hot—a lot less hot. The equatorial temperature is 10,300 K whereas the polar temperature is 15,400 K. Furthermore, since it takes Regulus only about 15.9 hours to make one rotation, the surface matter at Regulus's equator is moving at roughly 700,000 mph—that is 160 times faster than our Sun's and about 86 percent of the speed at which Regulus would start flying apart.

This great rotational speed also explains why the age of Regulus calculated from the appearance of its spectral lines did not agree with the ages derived for its pair of companion stars. We can now be assured that Regulus, like its companions, is only about 50 million years old. That is little more than one-hundredth the age of our Sun. Regulus was born rather long after the dinosaurs died.

These amazing findings derived from the McAllister et al. study I first reported in one of my columns in the April 2006 issue of *Sky & Telescope*. But I left out a final pair of findings that is every bit as remarkable and fascinat-

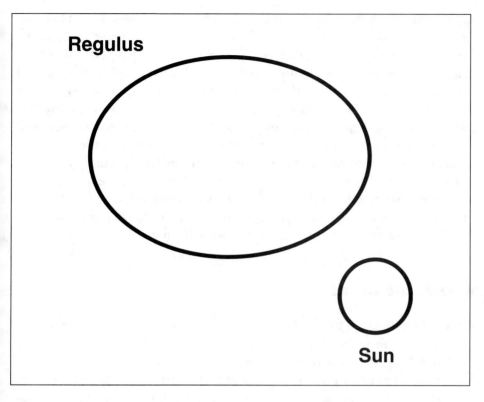

Regulus

Sun

Relative sizes and shapes of Regulus and the Sun.

ing. An astronomy writer had compared the oblate shape of Regulus to that of a pumpkin (good comparison—as long as we can live with the idea of a blue-white pumpkin). An editor of my *S&T* article added in a caption the idea that Regulus is "shaped not like a ball but like a spinning raindrop." That's a great comparison—as long as a person knows that, in real life, raindrops are not the cartoon teardrop shape (pointy top, rounded bottom) but like squashed—strongly oblate—spheres. I'm not sure if raindrops really do spin a lot (some ice crystals do—and thereby produce certain rare and unusual halo phenomena in the sky). One can even enjoy the idea that raindrops can perhaps appear slightly bluish—like Regulus. Of course, all analogies break down at some point. But, I digress. The additional information revealed by McAllister and colleagues is that the spin axis of Regulus is oriented almost exactly perpendicularly from our line of sight to the star—and almost exactly in the star's direction of motion. Regulus, in other words, looks to us like it is on its side and spinning like a bullet (or a "spiral pass" in American football, but with little if any wobble) flying almost precisely due west in our heavens.

Note that this "bullet" we call Regulus is not going through space at a speed particularly faster than other stars relative to the Sun. This is because Regulus, like most of the other stars, is part of the main stream of traffic orbiting the center of our galaxy with our Sun in the equatorial disk of the galaxy. It takes a star like Arcturus—plunging down through the equatorial disk of the galaxy on a highly inclined orbit—to speed about four times faster than Regulus's 29 km/sec space velocity. (By the way, compare this space velocity of Regulus with the star's 315 km/sec equatorial rotational velocity—the "bullet" spins more than ten times faster than it flies).

Is there some profoundly important reason why Regulus moves through space at a right angle to its direction of rotation? No one seems to know. Vega, now known to be fairly oblate and very fast-spinning (rotates at 92 percent of its breakup spin velocity!), is pointing its pole almost right at us—but is not traveling almost directly toward or away from us.

A Heart Oft-Visited

It's a little difficult to think of Regulus in complete isolation. The star commands Leo with its brightness and its central position in the heart of the imagined Lion. But we usually do observe it inextricably as part of Leo, or at least as the handle of Leo's conspicuous front part, the Sickle (Algieba, in the curve of the Sickle, is only about ⅔ magnitude dimmer than Regulus, as is Denebola, in the hour-and-a-half farther-east tail of Leo). With binoculars, there is the companion of Regulus—which in medium to large telescopes has a controversial color (is it orange-yellow or, as one great early observer had it, "seemingly steeped in indigo"?). Then there is the galaxy only about 20' almost due north of Regulus. This is the Leo I dwarf galaxy—at just 750,000 light-years away a member of our Local Group, yet so dim and so close to the glare of Regulus that it is most extremely difficult to glimpse visually. This galaxy was discovered photographically by A. G. Wilson with Palomar's giant 48-inch Schmidt telescope in 1950. Today, however, amateur astronomers are producing some marvelous photos showing both Regulus and Leo I together.

Nevertheless, it's not the constant neighbors of Regulus that draw our attention so much as its frequent visitors, many of which are spectacularly bright. I refer to the Moon and planets.

In my four decades of being knowledgeable enough to catch such sights, there have been literally hundreds of either good or great conjunctions of Moon and planets with Regulus. In 2007 and 2008, the best such events involve meetings of Regulus with Venus, Saturn, and the Moon—especially its meeting with the Moon and Saturn on the night of February 20–21, 2008.

That night was when the Moon was totally eclipsed for virtually all of the Americas, Europe, and Africa, and glowed a dimmed red with Saturn and Regulus flamed into full brightness only a few degrees to either side of it.

My most cherished memories that involve Regulus are those of it combining with the Moon and/or planets to create a scene of glory.

I've actually only caught once in my life, using a telescope, an occultation of Regulus by the Moon. And I was far too young to observe the incredibly rare occultation of Regulus by Venus on July 7, 1959.

But I do recall a magical morning when I was hoping to see a close Venus-Regulus conjunction before dawn. I had a tiny "window" of clear sky open in the overcast just where the brilliant planet and companionable Lion star were. An absolutely important point to make is that when a planet gets to even a few degrees from Regulus (many conjunctions are closer) the contrasting color of planet and star enhances wonderfully the hues of both. (There must be a contrast in every case, because Regulus is blue-white and there is no bright planet in our sky which is blue.) When that window opened in the clouds for me, it contained for my naked eye and binoculars a richly golden Venus next to a sapphire Regulus. And then, in the silence of that hour, I heard, at first as just a few soft notes so unexpected they were utterly strange—a sound that grew to a patter then to soft swish and rush. It was the sound of raindrops falling on the leaves of the trees and forest I stood near. I gazed on through the immaculately clear porthole or eyelet in which paired contrastingly colored heavenly lamps glowed while a gentle rain fell on me and all about me.

A few years later, I was still in the prime of young adulthood when I watched all the historic triple (three-event) conjunctions of Jupiter-Saturn and Jupiter-Mars in 1979 to 1981. I've discussed in my book *The Starry Room* these and the other interlinked planetary events that I think made those and neighboring years "the greatest years of the planets." Among the events were, twice, the only opportunities for centuries to see at one time all eight of our fellow planets. I say eight because the tally included Pluto, which—whether it stays demoted from planethood or not in the years to come—was visible with Mercury, Venus, Mars, Jupiter, Saturn, Uranus, and Neptune on those two occasions, one in February 1982 and one in January 1984. (At the first event, the Moon and the first four asteroids were also visible!).

The title of this chapter in *The Starry Room* is "All the Worlds in My Window"—which they were on one particular morning in January 1984. But, although I named the chapter after the all-planet events, I must say that the single most heartachingly beautiful and royally grand event for the eyes among the dozens in those great years was another one. And it was one that involved the star heart of the royal constellation beast—Regulus.

The evening was that of May 3, 1980. For months, Jupiter and Mars had played back and forth to either side of Regulus and had engaged in the first two meetings of their rare three-conjunction series. Now, at last, all came together. The sky was wonderfully clear, and the leaves were just coming on the trees. And, as twilight started to fade, there came into view, high in the south, the most splendid "trio" of celestial points of light I've ever seen. Jupiter burned at magnitude −2.2, Mars at +0.2, and Regulus at +1.3. And they were all located within a circle little more than $1\frac{1}{2}°$ wide. When full darkness fell, the beauteous bunching was a whole qualitative level or two of glory better than even the sum of its parts. The colors of all three heavenly lights—golden Jupiter, ruddy Mars, and blue Regulus—were enhanced by their proximity to one another. For hours, this stupendous triple gem traveled with the Lion across the heavens, before setting in the middle of the night. That evening, the heart of Leo the Lion, like the heart of me, its watcher, filled to the brim and overflowed with wonder.

APPENDIX A

The Brightest Stars: Position, Spectral Type, Apparent and Absolute Magnitude, and Distance

Stars are listed in order of apparent brightness.

Star	RA (2000)	Declination (2000)	Spectral Type	Apparent Magnitude	Absolute Magnitude	Distance (in light-years)
Sirius	06h 45m	−16.7°	A1V	−1.44	1.45	8.60
Canopus	06h 24m	−52.7°	F0Ib	−0.62	−5.53	310
Alpha Centauri	14h 40m	−60.8°	G2V+K1V	−0.27[a]	4.08	4.39
Arcturus	14h 16m	+19.2°	K2III	−0.05	−0.31	36.7
Vega	18h 37m	+38.8°	A0V	0.03	0.58	25.3
Capella	05h 17m	+46.0°	G5III+G0III	0.08	−0.48	42
Rigel	05h 15m	−8.2°	B8Ia	0.18	−6.69	770[b]
Procyon	07h 39m	+5.2°	F5IV-V	0.40	2.68	11.4
Achernar	01h 38m	−57.2°	B3V	0.45	−2.77	144
Betelgeuse	05h 55m	+7.4°	M2Ib	0.45[c]	−5.14[b]	430[b]
Beta Centauri	14h 04m	−60.4°	B1III	0.61	−5.42	530
Altair	09h 51m	+8.9°	A7V	0.76	2.20	26.7
Alpha Crucis	12h 27m	−63.1°	B0.5IV+B1V	0.76[a]	−4.19	320
Aldebaran	04h 36m	+16.5°	K5III	0.87	−0.63	65
Spica	13h 25m	−11.2°	B1V+B2V	0.98	−3.55	260
Antares	16h 29m	−26.4°	M1Ib+B4V	1.06[c]	−5.28	600[b]
Pollux	07h 45m	+28.0°	K0III	1.16	1.09	33.5
Fomalhaut	22h 58m	−29.6°	A3V	1.17	1.74	25.2
Beta Crucis	12h 48m	−59.7°	B0.5III	1.25	−3.92	350
Deneb	20h 41m	+45.3°	A2Ia	1.25	−8.73[b]	3,000[b]
Regulus	10h 08m	+12.0°	B7V	1.36	−0.52	78

[a] Combined magnitude of double star.
[b] Approximate. For Deneb, maximum values.
[c] Variable enough in brightness for changes to be noted visually.

APPENDIX B

The Brightest Stars: Spectral Type, Color Index, Color, and Surface Temperature

Star	Spectral Type	B-V Index	Color	Temperature (degrees Kelvin)
Sun	G2V	0.66	yellow	5,780
Sirius	A1V	0.01	blue-white	9,880
Canopus	F0Ib	0.16	white	7,800
Alpha Centauri	G2V + K1V	0.73	yellow	5,770 + 5,300
Arcturus	K2III	1.23	light orange	4,290
Vega	A0V	0.00	blue-white	9,600 (polar)/7,400 (equator)
Capella	G5III+G0III	0.80	yellow	4,940? + 5,700
Rigel	B8Ia	−0.03	blue-white	11,000
Procyon	F5IV-V	0.42	yellow-white	6,530
Achernar	B3V	−0.16	blue-white	14,500 – 19,300
Betelgeuse	M2Ib	1.84	deep orange	~3,600
Beta Centauri	B1III	−0.24	blue-white	~22,500 + ~22,500
Alpha Crucis	B0.5IV + B1V	−0.24	blue-white	~28,000 + ~26,000
Altair	A7V	0.22	slightly yellow-white	7,550
Aldebaran	K5III	1.54	orange	4,000
Spica	IV + B2V	−0.23	blue-white	2,400 + 18,500
Antares	I Ib + B4V	1.84	deep orange	3,600
Pollux	K0IIIb	1.00	light orange	4,770
Fomalhaut	3V	0.12	white	8,500
Beta Crucis	0.5III	0.24	blue-white	7,600
Deneb	2Ia	0.09	white	400
Regulus	7V	−0.10	blue-white	2,000

APPENDIX C

Midnight and 9:00 P.M. Culminations, Season of Prime Evening Visibility

Culmination is the reaching of the star's highest point, on the north–south meridian of the sky.

Star	Greek Letter Designation	Culmination (midnight/9:00 P.M.)	Season of Prime Evening Visibility
Sirius	Alpha Canis Majoris	January 1/February 16	Winter
Canopus	Alpha Carinae	December 27/February 11	Summer*
Alpha Centauri	Alpha Centauri	May 3/June 16	Late fall*
Arcturus	Alpha Boötis	April 27/June 10	Late spring
Vega	Alpha Lyrae	July 1/August 15	Summer
Capella	Alpha Aurigae	December 12/January 24	Winter
Rigel	Beta Orionis	December 12/January 24	Winter
Procyon	Alpha Canis Minoris	January 14/March 2	Late winter
Achernar	Alpha Eridani	October 20/November 30	Late spring*
Betelgeuse	Alpha Orionis	December 21/February 3	Winter
Beta Centauri	Beta Centauri	April 23/June 7	Late fall*
Alpha Crucis	Alpha Crucis	March 28/May 13	Fall*
Altair	Alpha Aquilae	July 18/September 3	Late summer
Aldebaran	Alpha Crucis	December 3/January 14	Early winter
Spica	Alpha Virginis	April 13/May 28	Late spring
Antares	Alpha Scorpii	May 30/July 14	Early summer
Pollux	Beta Geminorum	January 15/March 3	Late spring
Fomalhaut	Alpha Piscis Austrini	September 6/October 20	Early fall
Beta Crucis	Beta Crucis	April 2/May 18	Fall*
Deneb	Alpha Cygn	August 1/September 16	Late summer
Regulus	Alpha Leonis	February 19/April 9	Early spring

*Only visible south of 40° N; season of prime evening visibility is Southern Hemisphere season (reverse of North Hemisphere season).

APPENDIX D

Diameters and Masses of the Brightest Stars

Star	Spectral Type	Diameter (Sun = 1)	Mass (Sun = 1)
Sun	G2V	1.00	1.00
Sirius	A1V	1.75	2.12
Canopus	F0lb	65	8–9
Alpha Centauri	G2V + K1V	1.26	1.10
Arcturus	K2III	26	1.5
Vega	A0V	2.25 × 2.75	2.5
Capella	G5III + G0III	10 +10	2.7 + 2.6
Rigel	B8la	70	17*
Procyon	F5IV-V	2.1	1.4
Achernar	B3V	~7.1 x ~11.8	6–8
Betelgeuse	M2lb	~600	12* – 17*
Beta Centauri	B1III	8 + 8	10.7 +10.3
Alpha Crucis	B0.5IV + B1V	?/6.7 + ? + 5.5	14/10 + 13
Altair	A7V	~1.7 x ~1.8	1.7 – 1.8
Aldebaran	K5III	40	2.5
Spica	B1V + B2V	7.8 + 4.0	11 + 7
Antares	M1lb + B4V	~800+	15* – 18* + 7–8
Pollux	K0III	10	1.8
Fomalhaut	A3V	~1.5	
Beta Crucis	B0.5III	8.1	14
Deneb	A2la	~200	20*
Regulus	B7V	3.15 x 4.15	3.5

*Estimated original mass, some of which has been lost.

APPENDIX E

Motions of the Brightest Stars

Star	Parallax Amount (milli-")	Proper Motion ang(")	Proper Motion (degrees)	Proper Motion Direction	Velocity Radial* Space (km/sec)	Velocity Radical Space (km/sec)
Sirius	379	1.324	204	SSW	−8	19
Canopus	10	0.034	50	NE	21	25
Alpha Centauri	742	3.678	281	W/WNW	−23	33
Arcturus	89	2.281	209	SSW	−5	122
Vega	129	0.348	35	NE	−14	19
Capella	77	0.340	169	SSE/S	30	40
Rigel	4	0.004	236	SW/WSW	21	21
Procyon	286	1.248	214	SSW/SW	−3	21
Achernar	23	0.108	105	ESE	16	26
Betelgeuse	8	0.028	68	ENE	21	28
Beta Centauri	6	0.030	221	SW	6	33
Alpha Crucis A	10	0.030	236	SW/WSW	−11	20
Altair	194	0.662	54	NE/ENE	−26	30
Aldebaran	50	0.200	161	SSE	57	54
Spica	12	0.054	232	SW	1	20
Antares	5	0.024	197	SSW	−3	35
Pollux	97	0.629	265	W	3	31
Fomalhaut	130	0.373	116	ESE	7	15
Beta Crucis	9	0.042	246	WSW	16	30
Deneb	1	0.005	11	N/NNE	−5	9
Regulus	42	0.248	271	W	6	29

*Radial velocity in positive numbers is receding, in negative numbers is approaching.

APPENDIX F

The 200 Brightest Stars

To keep this data consistent with itself, some of the statistics disagree with the figures elsewhere in the book. For instance, Alpha Crucis is listed as slightly fainter, 0.77 instead of 0.76, placing it after Altair in the ranking. Also, this list gives the older figures for the distance and absolute magnitude of Beta Centauri.

Greek Name	Proper Name	RA for 2000	Declination for 2000	Spectral Classification	Visual Magnitude	Absolute Magnitude	Distance (in light-years)
Alpha Canis Majoris	Sirius	06h 45m	−16.7°	A1V	−1.44	1.45	9
Alpha Carinae	Canopus	06h 24m	−52.7°	F0Ib	−0.62	−5.53	310
Alpha Centauri	Rigil Kentaurus	14h 40m	−60.8°	G2V + K1V	−0.27	4.08	4
Alpha Boötes	Arcturus	14h 16m	+19.2°	K2III	−0.05	−0.31	37
Alpha Lyrae	Vega	18h 37m	+38.8°	A0V	0.03	0.58	25
Alpha Aurigae	Capella	05h 17m	+46.0°	G5III + G0III	0.08	−0.48	42
Beta Orionis	Rigel	05h 15m	−8.2°	B8Ia	0.18	−6.69	770
Alpha Canis Minoris	Procyon	07h 39m	+5.2°	F5IV-V	0.40	2.68	11
Alpha Eridani	Achernar	01h 38m	−57.2°	B3V	0.45	−2.77	144
Alpha Orionis	Betelgeuse	05h 55m	+7.4°	M2Ib	0.45	−5.14	430
Beta Centauri	Hadar	14h 04m	−60.4°	B1III	0.61	−5.42	530
Alpha Aquilae	Altair	19h 51m	+8.9°	A7V	0.76	2.20	17
Alpha Crucis	Acrux	12h 27m	−63.1°	B0.5IV + B1V	0.77	−4.19	320
Alpha Tauri	Aldebaran	04h 36m	+16.5°	K5III	0.87	−0.63	65
Alpha Virginis	Spica	13h 25m	−11.2°	B1V + B2V	0.98	−3.55	260
Alpha Scorpii	Antares	16h 29m	−26.4°	M1Ib + B4V	1.06	−5.28	600
Beta Geminorum	Pollux	07h 45m	+28.0°	K0III	1.16	1.09	34
Alpha Piscis Austrini	Fomalhaut	22h 58m	−29.6°	A3V	1.17	1.74	25
Beta Crucis	Mimosa	12h 48m	−59.7°	B0.5III	1.25	−3.92	350
Alpha Cygni	Deneb	20h 41m	+45.3°	A2Ia	1.25	−8.73	3,000
Alpha Leonis	Regulus	10h 08m	+12.0°	B7V	1.36	−0.52	78
Epsilon Canis Majoris	Adhara	06h 59m	−29.0°	B2II	1.50	−4.10	430
Alpha Geminorum	Castor	07h 35m	+31.9°	A1V + A2V	1.58	0.59	52

Greek Name	Proper Name	RA for 2000	Declination for 2000	Spectral Classification	Visual Magnitude	Absolute Magnitude	Distance (in light-years)
Gamma Crucis	Gacrux	12h 31m	−57.1°	M3.5III	1.59	−0.56	88
Lambda Scorpii	Shaula	17h 34m	−37.1°	B2IV	1.62	−5.05	700
Gamma Orionis	Bellatrix	05h 25m	+6.3°	B2III	1.64	−2.72	240
Beta Tauri	Elnath	05h 26m	+28.6°	B7III	1.65	−1.37	130
Beta Carinae	Miaplacidus	09h 13m	−69.7°	A2III	1.67	−0.99	111
Epsilon Orionis	Alnilam	05h 36m	−1.2°	B0Ia	1.69	−6.38	1,300
Alpha Gruis	Alnair	22h 08m	−47.0°	B7IV	1.73	−0.73	101
Zeta Orionis	Alnitak	05h 41m	−1.9°	O9.5Ib + B0III	1.74	−5.26	820
Gamma Velorum	Regor	08h 10m	−47.3°	WC8 + O9Ib	1.75	−5.31	840
Epsilon Ursae Majoris	Alioth	12h 54m	+56.0°	A0IV	1.76	−0.21	81
Alpha Persei	Mirfak	03h 24m	+49.9°	F5Ib	1.79	−4.50	590
Epsilon Sagittarii	Kaus Australis	18h 24m	−34.4°	B9.5III	1.79	−1.44	145
Alpha Ursae Majoris	Dubhe	11h 04m	+61.8°	K0III + F0V	1.81	−1.08	124
Delta Canis Majoris	Wezen	07h 08m	−26.4°	F8Ia	1.83	−6.87	1,800
Eta Ursae Majoris	Alkaid	13h 48m	+49.3°	B3V	1.85	−0.60	101
Epsilon Carinae	Avior	08h 23m	−59.5°	K3II + B2V	1.86	−4.58	630
Theta Scorpii	Sargas	17h 37m	−43.0°	FIII	1.86	−2.75	270
Beta Aurigae	Menkalinan	06h 00m	+44.9°	A2IV	1.90	−0.10	82
Alpha Trianguli Australis	Atria	16h 49m	−69.0°	K2Ib-II	1.91	−3.62	420
Gamma Geminorum	Alhena	06h 38m	+16.4°	A0IV	1.93	−0.60	105
Delta Velorum	Koo She	08h 45m	−54.7°	A0V	1.93	−0.01	80
Alpha Pavonis	Peacock	20h 26m	−56.7°	B0.5V + B2V	1.94	−1.81	180
Alpha Ursae Minoris	Polaris	02h 32m	+89.3°	F7Ib-II	1.97	−3.64	430
Beta Canis Majoris	Mirzam	06h 23m	−18.0°	B1III	1.98	−3.95	500
Alpha Hydrae	Alphard	09h 28m	−8.7°	K3II	1.99	−1.69	180
Alpha Arietis	Hamal	02h 07m	+23.5°	K2III	2.01	0.48	66
Gamma Leonis	Algieba	10h 20m	+19.8°	K0III + G7III	2.01	−0.92	126
Beta Ceti	Diphda	00h 44m	−18.0°	K0III	2.04	−0.30	96
Sigma Sagittarii	Nunki	18h 55m	−26.3°	B3V	2.05	−2.14	220
Theta Centauri	Menkent	14h 07m	−36.4°	K0III	2.06	0.70	61
Alpha Andromedae	Alpheratz	00h 08m	+29.1°	B9IV	2.07	−0.30	97
Beta Andromedae	Mirach	01h 10m	+35.6°	M0II	2.07	−1.86	200
Kappa Orionis	Saiph	05h 48m	−9.7°	B0.5III	2.07	−4.65	720
Beta Ursae Minoris	Kochab	14h 51m	+74.2°	K4III	2.07	−0.87	127

Continued

The 200 Brightest Stars *(continued)*

Greek Name	Proper Name	RA for 2000	Declination for 2000	Spectral Classification	Visual Magnitude	Absolute Magnitude	Distance (in light-years)
Gamma Gruis	Al Dhanab	22h 43m	−46.9°	M5III	2.07	−1.52	170
Alpha Ophiuchi	Rasalhague	17h 35m	+12.6°	A5III	2.08	1.30	47
Beta Persei	Algol	03h 08m	+41.0°	B8V + G5IV + A	2.09	−0.18	93
Gamma Andromedae	Almach	02h 04m	+42.3°	K3II + B8V + A0V	2.10	−3.08	360
Beta Leonis	Denebola	11h 49m	+14.6°	A3V	2.14	1.92	36
Gamma Cassiopeiae	Cih	00h 57m	+60.7°	B0IV	2.15	−4.22	610
Gamma Centauri	Muhlifain	12h 42m	−49.0°	A0III + A0III	2.20	−0.81	130
Zeta Puppis	Naos	08h 04m	−40.0°	O5Ia	2.21	−5.95	1,400
Iota Carinae	Aspidiske	09h 17m	−59.3°	A8Ib	2.21	−4.42	690
Alpha Coronae Borealis	Alphecca	15h 35m	+26.7°	A0V + G5V	2.22	0.42	75
Lambda Velorum	Suhail	09h 08m	−43.4°	K4Ib	2.23	−3.99	570
Zeta Ursae Majoris	Mizar	13h 24m	+54.9°	A2V + A2V + A1V	2.23	0.33	78
Gamma Cygni	Sadr	20h 22m	+40.3°	F8Ib	2.23	−6.12	1,500
Alpha Cassiopeiae	Schedar	00h 41m	+56.5°	K0II	2.24	−1.99	230
Gamma Draconis	Eltanin	17h 57m	+51.5°	K5III	2.24	−1.04	148
Delta Orionis	Mintaka	05h 32m	−0.3°	O9.5II + B2V	2.25	−4.99	920
Beta Cassiopeiae	Caph	00h 09m	+59.2°	F2III	2.28	1.17	55
Epsilon Centauri	Birdun	13h 40m	−53.5°	B1III	2.29	−3.02	380
Delta Scorpii	Dschubba	16h 00m	−22.6°	B0.5IV	2.29	−3.16	400
Epsilon Scorpii	Wei	16h 50m	−34.3°	K2.5III	2.29	0.78	65
Alpha Lupi	Men	14h 42m	−47.4°	B1.5III	2.30	−3.83	550
Eta Centauri		14h 36m	−42.2°	B1.5V	2.33	−2.55	310
Beta Ursae Majoris	Merak	11h 02m	+56.4°	A1V	2.34	0.41	79
Epsilon Boötis	Izar	14h 45m	+27.1°	K0II − III + A2V	2.35	−1.69	210
Epsilon Pegasi	Enif	21h 44m	+9.9°	K2Ib	2.38	−4.19	670
Theta Scorpii	Girtab	17h 42m	−39.0°	B1.5III	2.39	−3.38	460
Alpha Phoenicis	Ankaa	00h 26m	−42.3°	K0III	2.40	0.52	77
Gamma Ursae Majoris	Phecda	11h 54m	+53.7°	A0V	2.41	0.36	84
Eta Ophiuchi	Sabik	17h 10m	−15.7°	A1V + A3V	2.43	0.37	84
Beta Pegasi	Scheat	23h 04m	+28.1°	M2III	2.44	−1.49	200
Eta Canis Majoris	Aludra	07h 24m	−29.3°	B5Ia	2.45	−7.51	3,000
Alpha Cephei	Alderamin	21h 19m	+62.6°	A7IV	2.45	1.58	49
Kappa Velorum	Markeb	09h 22m	−55.0°	B2IV	2.47	−3.62	540
Epsilon Cygni	Gienah	20h 46m	+34.0°	K0III	2.48	0.76	72
Alpha Pegasi	Markab	23h 05m	+15.2°	B9IV	2.49	−0.67	140

Greek Name	Proper Name	RA for 2000	Declination for 2000	Spectral Classification	Visual Magnitude	Absolute Magnitude	Distance (in light-years)
Alpha Ceti	Menkar	03h 02m	+4.1°	M2III	2.54	−1.61	220
Zeta Ophiuchi	Han	16h 37m	−10.6°	O9.5V	2.54	−3.20	460
Zeta Centauri	Al Nair al Kentaurus	13h 56m	−47.3°	B2.5IV	2.55	−2.81	390
Delta Leonis	Zosma	11h 14m	+20.5°	A4V	2.56	1.32	58
Beta Scorpii	Graffias	16h 05m	−19.8°	B1V + B2V	2.56	−3.50	530
Alpha Leporis	Arneb	05h 33m	−17.8°	F0Ib	2.58	−5.40	1,300
Delta Centauri		12h 08m	−50.7°	B2IV	2.58	−2.84	400
Gamma Corvi	Gienah Ghurab	12h 16m	−17.5°	B8III	2.58	−0.94	165
Zeta Sagittarii	Ascella	19h 03m	−29.9°	A2IV + A4V	2.60	0.42	89
Beta Librae	Zubeneschamali	15h 17m	−9.4°	B8V	2.61	−0.84	160
Alpha Serpentis	Unukalhai	15h 44m	+6.4°	K2III	2.63	0.87	73
Beta Arietis	Sheratan	01h 55m	+20.8°	A5V	2.64	1.33	60
Alpha Librae	Zubenelgenubi	14h 51m	−16.0°	A3IV + F4IV	2.64	0.77	77
Alpha Columbae	Phact	05h 40m	−34.1°	B7IV	2.65	−1.93	270
Theta Aurigae		06h 00m	+37.2°	A0III + G2V	2.65	−0.98	170
Beta Corvi	Kraz	12h 34m	−23.4°	G5III	2.65	−0.51	140
Delta Cassiopeiae	Ruchbah	01h 26m	+60.2°	A5III	2.66	0.24	99
Eta Boötis	Muphrid	13h 55m	+18.4°	G0IV	2.68	2.41	37
Beta Lupi	Ke Kouan	14h 59m	−43.1°	B2III	2.68	−3.35	520
Iota Aurigae	Hassaleh	04h 57m	+33.2°	K3II	2.69	−3.29	510
Mu Velorum		10h 47m	−49.4°	G5III + G2V	2.69	−0.06	116
Alpha Muscae		12h 37m	−69.1°	B2V	2.69	−2.17	310
Upsilon Scorpii	Lesath	17h 31m	−37.3°	B2IV	2.70	−3.31	520
Pi Puppis		07h 17m	−37.1°	K4Ib	2.71	−4.92	1,100
Delta Sagittarii	Kaus Meridionalis	18h 21m	−29.8°	K2II	2.72	−2.14	310
Gamma Aquilae	Tarazed	19h 46m	+10.6°	K3II	2.72	−3.03	460
Delta Ophiuchi	Yed Prior	16h 14m	−3.7°	M1III	2.73	−0.86	170
Eta Draconis	Aldhibain	16h 24m	+61.5°	G8III	2.73	0.58	88
Theta Carinae		10h 43m	−64.4°	B0V	2.74	−2.91	440
Gamma Virginis	Porrima	12h 42m	−1.5°	F0V + F0V	2.74	2.38	39
Iota Orionis	Hatysa	05h 35m	−5.9°	O9III	2.75	−5.30	1,300
Iota Centauri		13h 21m	−36.7°	A2V	2.75	1.48	59
Beta Ophiuchi	Cebalrai	17h 43m	+4.6°	K2III	2.76	0.76	82
Beta Eridani	Kursa	05h 08m	−5.1°	A3III	2.78	0.60	89
Beta Herculis	Kornephoros	16h 30m	+21.5°	G7III	2.78	−0.50	150
Delta Crucis		12h 15m	−58.7°	B2IV	2.79	−2.45	360
Beta Draconis	Rastaban	17h 30m	+52.3°	G2II	2.79	−2.43	360

Continued

The 200 Brightest Stars *(continued)*

Greek Name	Proper Name	RA for 2000	Declination for 2000	Spectral Classification	Visual Magnitude	Absolute Magnitude	Distance (in light-years)
Alpha Canum Venaticorum	Cor Caroli	12h 56m	+38.3°	A0IV + F0V	2.80	0.16	110
Gamma Lupi		15h 35m	−41.2°	B2IV-V + B2IV-V	2.80	−3.40	570
Beta Leporis	Nihal	05h 28m	−20.8°	G5III	2.81	−0.63	160
Zeta Herculis	Rutilicus	16h 41m	+31.6°	F9IV + G7V	2.81	2.64	35
Beta Hydri		00h 26m	−77.3°	G2IV	2.82	3.45	24
Tau Scorpii		16h 36m	−28.2°	B0V	2.82	−2.78	430
Lambda Sagittarii	Kaus Borealis	18h 28m	−25.4°	K1III	2.82	0.95	77
Gamma Pegasi	Algenib	00h 13m	+15.2°	B2IV	2.83	−2.22	330
Rho Puppis	Turais	08h 08m	−24.3°	F6III	2.83	1.41	63
Beta Trianguli	Australis	15h 55m	−63.4°	F2IV	2.83	2.38	40
Zeta Persei		03h 54m	+31.9°	B1II + B8IV + A2V	2.84	−4.55	980
Beta Arae		17h 25m	−55.5°	K3Ib-II	2.84	−3.49	600
Alpha Arae	Choo	17h 32m	−49.9°	B2V	2.84	−1.51	240
Eta Tauri	Alcyone	03h 47m	+24.1°	B7III	2.85	−2.41	370
Epsilon Virginis	Vindemiatrix	13h 02m	+11.0°	G8III	2.85	0.37	102
Delta Capricorni	Deneb Algedi	21h 47m	−16.1°	A5V	2.85	2.49	39
Alpha Hydri		01h 59m	−61.6°	F0III	2.86	1.16	71
Delta Cygni		19h 45m	+45.1°	B9.5III + F1V	2.86	−0.74	170
Mu Geminorum	Tejat	06h 23m	+22.5°	M3III	2.87	−1.39	230
Gamma Trianguli	Australis	15h 19m	−68.7°	A1III	2.87	−0.87	180
Alpha Tucanae		22h 19m	−60.3°	K3III	2.87	−1.05	200
Theta Eridani	Acamar	02h 58m	−40.3°	A4III + A1V	2.88	−0.59	160
Pi Sagittarii	Albaldah	19h 10m	−21.0°	F2II	2.88	−2.77	440
Beta Canis Minoris	Gomeisa	07h 27m	+08.3°	B8V	2.89	−0.70	170
Pi Scorpii		15h 59m	−26.1°	B1V + B2V	2.89	−2.85	460
Epsilon Persei		03h 58m	+40.0°	B0.5V + A2V	2.90	−3.19	540
Sigma Scorpii	Alniyat	16h 21m	−25.6°	B1III	2.90	−3.86	730
Beta Cygni	Albireo	19h 31m	+28.0°	K3II + B8V + B9V	2.90	−2.31	390
Beta Aquarii	Sadalsuud	21h 32m	−05.6°	G0Ib	2.90	−3.47	610
Gamma Persei		03h 05m	+53.5°	G8III + A2V	2.91	−1.57	260
Upsilon Carinae		09h 47m	−65.1°	A7Ib + B7III	2.92	−5.56	1,600
Eta Pegasi	Matar	22h 43m	+30.2°	G2II-III + F0V	2.93	−1.16	215
Tau Puppis		06h 50m	−50.6°	K1III	2.94	−0.80	185
Delta Corvi	Algorel	12h 30m	−16.5°	B9.5V	2.94	0.79	88
Alpha Aquarii	Sadalmelik	22h 06m	−00.3°	G2Ib	2.95	−3.88	760

Greek Name	Proper Name	RA for 2000	Declination for 2000	Spectral Classification	Visual Magnitude	Absolute Magnitude	Distance (in light-years)
Gamma Eridani	Zaurak	03h 58m	−13.5°	MIIII	2.97	−1.19	220
Zeta Tauri	Alheka	05h 38m	+21.1°	B4III	2.97	−2.56	420
Epsilon Leonis	Ras Elased Australis	09h 46m	+23.8°	GIII	2.97	−1.46	250
Gamma Sagittarii	Alnasl	18h 06m	−30.4°	K0III	2.98	0.63	96
Gamma Hydrae		13h 19m	−23.2°	G8III	2.99	−0.05	132
Iota Scorpii		17h 48m	−40.1°	F2Ia	2.99	−5.71	1,800
Zeta Aquilae	Deneb el Okab	19h 05m	+13.9°	A0V	2.99	0.96	83
Beta Trianguli		02h 10m	+35.0°	A5III	3.00	0.09	124
Psi Ursae Majoris		11h 10m	+44.5°	K1III	3.00	−0.27	147
Gamma Ursae Minoris	Pherkad Major	15h 21m	+71.8°	A3II	3.00	−2.84	480
Mu Scorpii		16h 52m	−38.0°	B1.5V + B6.5V	3.00	−4.01	820
Gamma Gruis		21h 54m	−37.4°	B8III	3.00	−0.97	205
Delta Persei		03h 43m	+47.8°	B5III	3.01	−3.04	530
Zeta Canis Majoris	Furad	06h 20m	−30.1°	B2.5V	3.02	−2.05	340
Omicron Canis Majoris		07h 03m	−23.8°	B3Ia	3.02	−6.46	2,600
Epsilon Corvi	Minkar	12h 10m	−22.6°	K2II	3.02	−1.82	300
Epsilon Aurigae	Almaaz	05h 02m	+43.8°	F0Ia	3.03	−5.95	2,000
Beta Muscae		12h 46m	−68.1°	B2V + B3V	3.04	−1.86	310
Gamma Boötis	Seginus	14h 32m	+38.3°	A7III	3.04	0.96	85
Beta Capricorni	Dabih	20h 21m	−14.8°	G5II + A0V	3.05	−2.07	340
Epsilon Geminorum	Mebsuta	06h 44m	+25.1°	G8Ib	3.06	−4.15	900
Mu Ursae Majoris	Tania Australis	10h 22m	+41.5°	M0III	3.06	−1.35	250
Delta Draconis	Tais	19h 13m	+67.7°	G9III	3.07	0.63	100
Eta Sagittarii		18h 18m	−36.8°	M3.5III	3.10	−0.20	149
Zeta Hydrae		08h 55m	+05.9°	G9III	3.11	−0.21	150
Nu Hydrae		10h 50m	−16.2°	K2III	3.11	−0.03	139
Lambda Centauri		11h 36m	−63.0°	B9III	3.11	−2.39	410
Alpha Indi		20h 38m	−47.3°	K0III	3.11	0.65	101
Beta Columbae	Wazn	05h 51m	−35.8°	K2III	3.12	1.02	86
Iota Ursae Majoris	Talitha	08h 59m	+48.0°	A7IV	3.12	2.29	48
Zeta Arae		16h 59m	−56.0°	K3II	3.12	−3.11	570
Delta Herculis	Sarin	17h 15m	+24.8°	A3IV	3.12	1.21	78
Kappa Centauri	Ke Kwan	14h 59m	−42.1°	B2IV	3.13	−2.96	540
Alpha Lyncis		09h 21m	+34.4°	K7III	3.14	−1.02	220
N Velorum		09h 31m	−57.0°	K5III	3.16	−1.15	240
Pi Herculis		17h 15m	+36.8°	K3II	3.16	2.10	370

GLOSSARY

absolute magnitude The magnitude of brightness a star would have if seen at a standard distance of 10 parsecs (about 32.6 light-years).

altazimuth system A system for indicating positions in the sky, using altitude and azimuth as vertical and horizontal measure.

altitude Apparent angular height in the sky (vertical measurement in the altazimuth system).

apparent magnitude The magnitude of brightness a star appears to have in our sky.

asterism A pattern of stars in the sky that is not an official constellation.

atmospheric extinction The dimming of the light of celestial objects due to absorption and scattering by Earth's atmosphere.

averted vision A technique of looking slightly to the side of a faint object to increase its visibility by allowing its light to fall on the parts of the eye's retina most sensitive to light.

azimuth The horizontal measure around the sky in the altazimuth system.

binary star A double star system in which the members are believed to be orbiting around each other (that is, around a common center of gravity).

black hole An object, thought to be the result of a massive star's collapse, whose gravity has become so strong as to prevent even light (and other electromagnetic radiations) from escaping from it.

blue giant A massive, large, and extremely luminous star with a very high surface temperature that causes it to appear bluish-white to the eye.

celestial objects Bodies in outer space beyond Earth's atmosphere.

celestial sphere The imaginary sphere surrounding Earth whose inner surface is all the sky containing astronomical objects both above and below one's horizon.

Cepheid A type of pulsating variable star whose precisely regular brightness variations can be employed to estimate the star's distance by use of the "period-luminosity relation." (If you know the period of the Cepheid's brightness variations, you know its luminosity—true brightness—and can therefore compare this to its apparent brightness to determine distance.)

circumpolar Close enough to one of the celestial poles so as to never rise or set but to instead circle around the pole above the horizon.

conjunction Strictly speaking, the arrangement when one celestial object moves to have the same right ascension or "ecliptic longitude" (passes due north or due south of a second object in a celestial coordinate system) of another. More loosely, any temporary pairing or gathering of celestial objects that is considered close.

constellation An official pattern of stars or, more strictly, the officially demarcated section of sky in which that pattern lies.

culminate Reach the north–south meridian of the sky, typically achieving highest point above horizon.

dark adaptation The increase in the sensitivity of our eyes to dim light that occurs when they are kept away from bright light for a while.

dark nebula A nebula that does not shine by either emitted or reflected light and is therefore visible only in silhouette against a more distant bright nebula or starry background.

declination North–south measure in the equatorial system of celestial coordinates, corresponding to latitude on Earth.

deep-sky object An object beyond our solar system, though the term is usually not applied to individual stars or double- and multiple-star systems but rather to star clusters, nebulae, and galaxies.

diffuse nebula A luminous nebula that shines from reflecting the light of nearby stars ("reflection nebula") or is heated enough by very hot stars to glow on its own ("emission nebula").

diurnal motion The apparent motion of celestial objects caused by the Earth's rotation.

double star A star that, upon closer or more sophisticated examination, turns out to consist of two or more component stars.

eclipse The hiding or dimming of one object by another object or by another object's shadow.

eclipsing binary A type of variable star in which one component star of a double star eclipses the other, or both alternately eclipse each other, causing the variations in brightness.

ecliptic   The apparent path of the Sun through the zodiac constellations, which is really the projection of Earth's orbit in the sky.

equatorial system   A system for indicating positions in the heavens using right ascension (corresponding to longitude on Earth) and declination (corresponding to latitude on Earth).

full-cutoff fixture   A light fixture that emits light entirely below the horizontal, eliminating directly produced skyglow and reducing light pollution.

galaxy   An immense congregation of typically billions of stars forming a system of spiral, elliptical, or irregular shape.

galactic cluster   See "open cluster."

globular cluster   A kind of star cluster consisting of tens of thousands up to a few million stars arranged in a roughly spherical shape.

horizon   The boundary line between sky and land or sea.

H-R diagram (Hertzsprung-Russell diagram)   A tremendously revealing diagram that plots the true brightness (in terms of absolute magnitude or luminosity) of stars versus their spectral class or surface temperature.

light pollution   Excessive or misdirected artificial outdoor lighting.

light-year   The distance that light, fastest thing in the universe, travels in the course of one year.

long-period variable   A major kind of variable star, in which the period of brightness variations is months or years long and the range of the variations typically great.

luminosity   The true brightness of a star, independent of its distance from us, measured in units of the Sun's true brightness (for instance, a star twice as luminous as the Sun has a luminosity of 2).

magnitude   A measure of brightness in astronomy, in which an object 100 times brighter than another is exactly 5 magnitudes brighter. The brighter the object, the lower the magnitude figure (e.g., a 1st-magnitude star is brighter than a 2nd-magnitude star), with negative magnitudes for the very brightest objects of all.

meridian   The line in the sky that passes from the due south horizon to the zenith onward to the due north horizon.

Messier objects (also known as **M-objects**)   One of slightly more than one hundred deep-sky objects cataloged in the eighteenth century by the French astronomer Charles Messier.

Milky Way   The great spiral galaxy that we live in, and also the night sky's band of glow from the combined light of innumerable distant stars in the galaxy's equatorial plane.

multiple star   A star system consisting of more than two stars (although "double star" is often used as the umbrella term for systems of two, three, four, etc., stars).

nebula   A vast cloud of dust and gas in interstellar space. Different types include "diffuse nebulae" (which include "emisssion nebulae" and "reflection nebulae"), "planetary nebulae," and "dark nebulae."

neutron star   The collapsed, ultra-dense core of a star left after a supernova formed from an original star not massive enough to collapse all the way into becoming a "black hole."

nova   An exploding (and therefore briefly very much brightened) star that loses a small fraction of its mass in the outburst, which may often arise from interactions between stars in double-star systems.

occultation   The hiding of one celestial object by another that is usually much larger (in a few cases, where the two bodies are not greatly different in apparent size, a hiding can be called an "eclipse").

open cluster (also called "galactic cluster")   This major kind of star cluster consists of a usually irregular shape and includes typically dozens of or a few hundred stars.

optical double   A double star in which the two components are not truly related, one object being much farther away and just happening to lie on nearly the same line of sight as seen from Earth.

PA See "position angle."

parallax   The change in a star's position caused by our change in viewpoint (usually our change from one side of Earth's orbit to the other).

parsec   A "parallax-second," the distance at which the view from opposite sides of Earth's orbit would cause an object to have an apparent position change (a "parallax") of 1 arc-second (1 parsec is equal to about 3.26 years).

planetary nebula   A cloud of gas and dust cast off by a hot, small, dying white dwarf star. (The name comes from the passing resemblance of some of these blue or green nebulae to the planets Uranus and Neptune seen through a telescope.)

position angle (PA)   The direction angle of a companion star in relation to its primary in a double-star system.

precession   A slight wobble in the rotational axis of Earth, caused by the pulls of the other solar system bodies, and resulting in slow changes of the direction of the north celestial pole and other positions in the heavens.

proper motion The motion of a star relative to the Sun as projected on the celestial sphere—in other words, the change in a star's position on the celestial sphere produced by the component of its "space velocity" (motion through space) that is transverse (neither toward nor away from us).

pulsar A type of neutron star oriented toward Earth in such a way that we get to observe regular, brief pulses of radio waves, light, or sometimes other electromagnetic wavelengths released from gaps in the exploding star's magnetic field near its poles.

quasar An incredibly powerful source of light and other electromagnetic wavelengths, which may be a kind of intense core of a galaxy.

RA See "right ascension."

red giant A huge star of extremely low density that radiates mostly in the red due to its relatively low surface temperature.

red dwarf A small star far less massive than the Sun, which radiates relatively little light and mostly in the red due to its comparatively low surface temperature.

revolution The orbiting of one celestial body around another (Earth's revolution period around the Sun is one year).

right ascension (RA) The west–east measure in the equatorial system of celestial coordinates, corresponding to longitude on Earth (though expressed somewhat differently than longitude: in hours of right ascension from 0 to 24).

rotation The spinning of a celestial object (Earth's rotation period is one day).

"seeing" Sharpness of astronomical images as a function of turbulence in Earth's atmosphere.

sidereal time Time measured by the passage of the stars around the sky, without reference to the Sun (the "sidereal day" is about 4 minutes shorter than the "solar day").

skyglow The component of "light pollution" (excessive and misdirected artificial outdoor lighting) that goes up into the sky. Around fairly large cities, it is visible for dozens of miles.

SNR See "supernova remnant."

solar system The whole collection of planets, moons, asteroids, comets, and meteoroids orbiting the Sun or another star, under the Sun's or other star's gravitational influence.

spectral class The subcategory of spectral type in which a star belongs—for instance, O3, G2, M5, in which the letter denotes the type and the number indicates the class.

spectral type The category—ranging from O (the hottest normal stars) to M (the coolest normal stars), with some alternative unusual types—into which a star is placed according to the appearance of its light's chemical spectrum.

star A massive self-luminous ball of gas producing energy by nuclear fusion in a dense core (also called a "sun").

star cluster A grouping of anywhere from a few to several million stars traveling through space relatively close together but not closely enough to be considered a multiple star.

supernova The much more powerful kind of star explosion and brightening in which a large part of a star's mass is lost and the star's core may become a neutron star or black hole.

supernova remnant (SNR) The cloud of material ejected by a supernova, sometimes visible for many thousands of years after.

transparency The degree to which Earth's atmosphere is capable of letting celestial light pass through it. (How clear of dust and water vapor is the atmosphere over you tonight?)

Universal time (UT) Time system for dating astronomical events, corresponding to the local time at Greenwich, England, on the 0° meridian of longitude on Earth (to obtain Universal time from your current local standard time, subtract 5 hours from Eastern Standard Time, 4 hours from Central Standard Time, and so on).

UT See "Universal Time."

variable star A star that, for one of a number of possible reasons, undergoes changes in its brightness.

white dwarf A hot, extremely small but fairly massive and therefore very dense star that represents the last luminous stage in the lives of many stars.

zenith The overhead point in the sky.

zodiac The circle of constellations through which the Sun passes during the course of the year.

SOURCES

As I stated in the Introduction of this book, there don't seem to be any other books devoted centrally to the brightest stars since 1907, the date when Martha Evans Martin's *The Friendly Stars* was originally published. On the other hand, a great number of books have been written that contain considerable information about the brightest stars within their wider-ranging coverage of stars or astronomy. In my comments on some of the books (and Web sites) that follow, the words in italics are my descriptions of what the books provide specifically on the topic of the brightest stars. If no italics appear, the book or Web sites may be simply a good source for background information on stars in general, or what it offers about the brightest stars is fairly obvious from its title (for instance, the book on double stars has specific information on the brightest stars that are double or multiple systems).

BOOKS

Allen, Richard Hinckley. *Star Names: Their Lore and Meaning.* New York: Dover Publications, 1963. Reprint of the classic 1899 edition. Allen has been faulted for his less than expert knowledge of Arabic but this work is by far the most comprehensive ever produced on star names and their lore. *It contains countless gems of lore and poetic quotes about individual stars, including many pages about the brightest stars.*

Burnham, Jr., Robert. *Burnham's Celestial Handbook* (3 volumes). New York: Dover Publications, 1978. The classic and nearly comprehensive guide to deep-sky objects visible through telescopes up to about 10-inch aperture. Some of the information has become outdated but the combination of vivid descriptions, useful tables, and inspired writing about lore and science has not been surpassed. *Burnham provides a verbal entry about the facts of every star that is magnitude 3.5 or brighter! It's too bad that a lot of his information about individual stars is now outdated.*

Houston, Walter Scott. *Deep-Sky Wonders.* Cambridge, Mass.: Sky Publishing Corporation, 1999. Collection of essays from the great Walter Scott Houston's seminal and long-running column about deep-sky objects. *Includes some insightful comments about a few of the brightest stars.*

Jobes, Gertrude and James. *Outer Space: Myths, Name Meanings, Calendars.* New York: Scarecrow Press, 1964. One of the few most interesting additions to Allen on the subject of starlore. The Jobeses' great expertise on folklore is partially offset by their lack of knowledge about astronomy and the book is riddled with typos—but it's worth having for some of the unique lore it offers and for Gertrude Jobes's superb storytelling and delightful poems. *Lots of lore about the brightest stars even if much of it is derived from Allen.*

Kaler, James R. *The Cambridge Encyclopedia of Stars.* London and New York: Cambridge University Press, 2006. I've not yet had a chance to read this book! But Kaler's "Stars" Web site (q.v.) and his many earlier books and articles have distinguished him as the premier writer of popular level books about the science of stars. So this latest book is likely to be a work of great value and distinction.

Kaler, James R. *The Hundred Greatest Stars.* New York: Copernicus Books, 2002. This is another Kaler book of distinction that I haven't read! This time, the reason is that I didn't want to be unduly influenced in the writing of the book you are holding in your hands (which I conceived of many years ago but have done most of the writing of in 2005–2007). This was my decision even though my book's focus is inherently different (Kaler's "greatest stars" list, interesting in its own right, apparently contains many stars that are not bright, some not even potentially visible to advanced amateur astronomers—perhaps because his emphasis is on the astrophysical?). Now that my own book is finished, good reader, I and you can seek out and read this almost certainly excellent book of Kaler's!

Kunitzsch, Paul, and Tim Smart. *A Dictionary of Modern Star Names*, 2nd rev. edition. Cambridge, Mass.: Sky Publishing Corporation, 2006. This small book is a guide to 254 star names and their derivations by our greatest contemporary expert on the subject, a German expert on Arabic and other Semitic languages, Paul Kunitzsch (Tim Smart has translated Kunitzsch's writing into English). *Invaulable discussions of the names of all the brightest stars.*

Martin, Martha Evans, and Donald Menzel. *The Friendly Stars.* New York: Dover Publications, 1963. Revised edition (revised by Menzel) of Martin's 1907 classic. *This book provides only very basic scientific information—but no one has ever written a more charming and heartfelt guide to the naked-eye observation and appreciation of the brightest stars.*

Mullaney, James. *Celestial Harvest.* New York: Dover Publications, 2002. One of the world's most veteran deep-sky observers offers wonderful descriptions, along with data, about more than three hundred of the finest deep-sky objects. *Descriptions of all of the brightest stars visible from midnorthern latitudes are included.*

————. *Double and Multiple Stars and How to Observe Them.* London: Springer, 2005. Excellent, spirited guide by probably the world's most experienced double-star observer.

Olcott, William Tyler. *Star Lore of All Ages.* New York: Dover Publications, 2005. This is a reprint of a 1911 classic, and includes a new foreword by the author of the book you are now holding. Olcott was arguably the greatest amateur astronomy writer of the first half of the twentieth century (He is best known for his *Field Book of the Skies.*) In *Star Lore of All Ages* he provides mostly constellation lore but *also much lore about the brightest stars, some of it not found in R. H. Allen—and all in Olcott's wonderful writing style.*

Ottewell, Guy. *The Astronomical Companion.* Universal Workshop (www.universalworkshop.com), 16th printing, with revisions, 2000. Ottewell's 72-page atlas-size guide to astronomical topics—eclipses, lunar phases and motions, types of stars, stellar evolution, etc.—which are year-independent (see his annual *Astronomical Calendar* discussed below, for astronomical phenomena that are year-dependent). This book includes dozens of huge and unique diagrams. *It contains a tremendous amount of information about stars—their brightnesses, distances, varieties, lives, and deaths—with some particular comments on the brightest stars. Of special interest is Ottewell's giant map of star names, followed by his translations of them all.*

Schaaf, Fred. *40 Nights to Knowing the Sky.* New York: Henry Holt and Company, 1998. Comprehensive introduction to the workings and appearances of the Moon, Sun, planets, stars, and deep-sky objects on the celestial sphere and in the universe.

————. *Seeing the Sky* and *Seeing the Deep Sky.* New York: John Wiley & Sons, 1990, 1992. Two books that offer observing projects focusing on naked-eye astronomical observations (*Seeing the Sky*) and telescopic observations of objects beyond the solar system (*Seeing the Deep Sky*). *There are many projects in each book concerning the stars, including a number about the brightest stars.*

STAR ATLASES, SKY SIMULATION SOFTWARE, PLANISPHERES

Star atlases range from those displaying only stars down to 5th or 6th magnitude (a few thousand stars) to *The Millennium Star Atlas* (which has well over 1 million stars, all stars brighter than magnitude 11.0). A fine selection of these atlases are described in the Sky Publishing Corporation catalog (see Web address for *Sky & Telescope* magazine).

An amazingly concise and informative little star atlas/handbook is E. Karkoschka's The *Observer's Sky Atlas* (New York: Springer, 2nd edition, 1999).

Various sky simulation software programs offer computer access to maps of even dimmer stars. Descriptions of many of these can also be found in the Sky Publishing catalog, and elsewhere (catalogs of the Astronomical Society of the Pacific and the Orion Telescope and Binocular Center, for instance).

If you need star-finding information that you can carry around and out to your observing site easily, try an old-fashioned planisphere, a handheld rotating wheel of cardboard or plastic that displays the starry sky for any time of any night. A superb all-around planisphere is the *Precision Planet and Star Locator,* produced by David Kennedal and distributed in North America by Sky Publishing Corporation.

PERIODICALS, ALMANACS, ANNUAL GUIDES

Astronomical Calendar. 82 atlas-size pages filled with original diagrams and rich text about the year's celestial events (includes several sections and month-by-month "Observers' Highlights" by the author of the book you are holding). By Guy Ottewell at Universal Workshop (www.universalworkshop.com). *Includes a section called "Deep-Sky Profiles," which sometimes has essays on one of the brightest stars by the author of this book.*

Astronomy Magazine www.astronomy.com. One of the two largest and most popular astronomy magazines (*Sky & Telescope* being the other).

Mercury. Fine magazine published by Astronomical Society of the Pacific. www.astrosociety.org.

Observer's Handbook (of Royal Astronomical Society of Canada). www.rasc.ca/handbook. Excellent annual guide with very extensive technical information for observers, written by a variety of experts. *Authoritative and up-to-date statistics and discussion about various kinds of stars, including the brightest stars in particular.*

Sky & Telescope www.skytonight.com. Most enduring of astronomy magazines. *Includes Sun, Moon, and planets column and star column by the author of this book.*

Sky Calendar. Abrams Planetarium, Michigan State University, East Lansing, MI 48824, (517) 355-4676 (see also "Skywatcher's Diary" at www.pa.msu.edu/abrams/diary.html). For just $11 a year, you get a two-sided sheet for each month with a superb basic star map on one side and an informative calendar with sky scene for each night on the other side.

SkyNews. www.skynewsmagazine.com. Canadian magazine much shorter than *Sky & Telescope* and *Astronomy* and published only bimonthly but very beautifully illustrated and offering some excellent observational articles.

ONLINE SOURCES

American Association of Variable Star Observers (AAVSO). www.aavso.org.

AstroAlerts. Available from the Sky & Telescope Web site (www.skytonight.com), this is a free service that sends the subscriber e-mail notices about urgently time-sensitive astronomical events that are in progress. A person can susbscribe to all the AstroAlerts or just to those that pertain to particular kinds of events (*such as novae, supernovae, and other unusual variable stars activity*).

Astronomical League. www.mcs.net/~bstevens/al. Confederation of more than two hundred amateur astronomy clubs in the United States.

Astronomical Society of the Pacific. www.astrosociety.org. This educational organization, more than a century old, offers a catalog with a wide variety of excellent books, posters, slide sets, software, CD-ROMs, etc.

Astronomy. www.astronomy.com. Web site of *Astronomy* magazine.

The Brighest Stars. www.atlasoftheuniverse.com/stars.html. *This is a beautifully presented (in part, color-coded) list of all the stars brighter than magnitude 3.55. It is based on Hipparcos data but without a few of the recent important refinements of distances (such as the lesser distance of Beta Centauri). Still, this is an impressive production. And even more impressive is Web site creator Richard Powell's suite of other, linked pages showing 3-D maps of the universe working outward from the closest stellar neighborhood of the Sun.*

Clear Sky Clock. www.cleardarksky.com. Attila Danko's invaluable predictions of sky conditions (cloud-cover, transparency, "seeing," and much more) for the next two days at innumerable sites in North America.

Hubble Heritage. http://heritage.stsci.edu. Popular access to Hubble Space Telescope images.

International Dark-Sky Association (IDA). www.darksky.org. Central bureau for information on light pollution and how to combat it, technologically and legally.

Interstellar Database. www.stellar-database.com. *Roger Wilcox's site, providing detailed statistics about many stars, though mostly those within 75 light-years of Earth (most interesting are the figures one can access here about the distances of neighbor stars as seen from a particular star).*

International Occultation Timing Association (IOTA). www.lunar-occultations.com/iota.

Sky & Telescope. www.skytonight.com. Web site of *Sky & Telescope* magazine and Sky Publishing Corporation.

Sol Station. www.solstation.com. A Web site featuring *excellent compilations of information about many stars (though mostly nearby ones), with many links to scientific papers.*

Space Weather Bureau. www.spaceweather.com. Latest predictions, information, and images about aurora and solar activity—and about a wonderful assortment of other astronomical objects and events. Great all-around astronomy Web site (*sometimes includes stunning images and information about the stars*).

Stars. www.astro.uiuc.edu/~kaler/sow/sowlist.html. *James Kaler's masterful Web site has concise, informative profiles—primarily about the astrophsysical aspects—of several hundred individual stars, with a new "Star of the Week" added each week. I have made extensive use of this in the research for this book of mine. Note also Kaler's suite of other astronomical pages linked to this one.*

ILLUSTRATION CREDITS

Page 8, © Bernhard Hubl; pp. 14, 16, 20, 22, © 2007 Abrams Planetarium; pp. 31, 43, 67, 75, 95, 115, 134, 146, 156, 160, 172, 181, 194, 217, 231, 247, 253, © Doug Myers; p. 40, © USAF; pp. 53, 56, 86, 182, 238, © NASA and STScI; pp. 65, 90, 96, 120, 221, © Guy Ottewell; pp. 105, 122, 137, 143, 207, © Roger Sinnott/*Sky & Telescope*; pp. 107, 129, 140, 149, 177, 200, 211, 218, 250, © Johannes Bayer's *Uranometria* (1603); p. 118, © ESO; pp. 147, 198, © Casey Reed/*Sky & Telescope*; p. 163, © Camille Flammarion's *Les Etoiles*; pp. 164, 205, © NASA/JPL; p. 166, © *Middleton Celestial Atlas* (1843); pp. 191, 227, 245, © William Tyler Olcott's *Field Book of the Skies*; p. 192, © Proctor's *Easy Star Lesson* (1883); p. 197, © John Monnier, University of Michigan); pp. 188, 243, © Akira Fujii; p. 203, © Steve Albers

INDEX

Page numbers in *italics* refer to illustrations.